Genders and Sexualities in History

Series Editors: John H. Arnold, Joanna Bourke and Sean Brady

Palgrave Macmillan's series, *Genders and Sexualities in History*, aims to accommodate and foster new approaches to historical research in the fields of genders and sexualities. The series will promote world-class scholarship that concentrates upon the interconnected themes of genders, sexualities, religions/religiosity, civil society, class formations, politics and war.

Historical studies of gender and sexuality have often been treated as disconnected fields, while in recent years historical analyses in these two areas have synthesized, creating new departures in historiography. By linking genders and sexualities with questions of religion, civil society, politics and the contexts of war and conflict, this series will reflect recent developments in scholarship, moving away from the previously dominant and narrow histories of science, scientific thought and legal processes. The result brings together scholarship from contemporary, modern, early modern, medieval, classical and non-Western history to provide a diachronic forum for scholarship that incorporates new approaches to genders and sexualities in history.

In the last few decades, the history of masculinity has drawn considerable academic interest. For the Middle Ages, the topic comes with a particular and immediate complication: 'men' are infrequently found as a catch-all category, being more frequently subdivided by status, or by role – those who fight, those who labor and those who pray. *Negotiating Clerical Identities: Priests, Monks and Masculinity in the Middle Ages* is the first volume of essays to focus specifically on clerical masculinity, and the collection forms part of a somewhat revisionary second wave of study in this area. The contributions discuss a variety of figures, texts and settings, drawn from across Europe in the central and later Middle Ages. Across this broad landscape, certain shared themes emerge, from the complications of chastity, and the interlacing of spiritual and warrior identities, to the potential tensions between prescription and lived experience. Throughout the book, we are reminded that neither 'sexuality' nor 'gender' can stand as discrete categories of analysis: what constitutes 'sexuality' in a particular historical period, for a particular category of person, is bound up with ideas about gender roles, which are themselves part and parcel of other cultural, social and indeed political forces. *Negotiating Clerical Identities* is essential reading for medievalists; it will also speak to a wider historical readership, in its framing of the interpretative and methodological issues involved in the study of men.

Titles include:

Christopher E. Forth and Elinor Accampo (*editors*)
CONFRONTING MODERNITY IN FIN-DE-SIÈCLE FRANCE
Bodies, Minds and Gender

Dagmar Herzog (*editor*)
BRUTALITY AND DESIRE
War and Sexuality in Europe's Twentieth Century

Jessica Meyer
MEN OF WAR
Masculinity and the First World War in Britain

Jennifer D. Thibodeaux (*editor*)
NEGOTIATING CLERICAL IDENTITIES
Priests, Monks and Masculinity in the Middle Ages

Hester Vaizey
SURVIVING HITLER'S WAR
Family Life in Germany 1939–48

Genders and Sexualities in History Series
Series Standing Order 978–0–230–55185–5 Hardback
978–0–230–55186–2 Paperback
(*outside North America only*)

You can receive future titles in this series as they are published by placing a standing order. Please contact your bookseller or, in case of difficulty, write to us at the address below with your name and address, the title of the series and the ISBN quoted above.

Customer Services Department, Macmillan Distribution Ltd, Houndmills, Basingstoke, Hampshire RG21 6XS, England

Negotiating Clerical Identities

Priests, Monks and Masculinity in the Middle Ages

Edited by

Jennifer D. Thibodeaux

Associate Professor of History, University of Wisconsin-Whitewater, USA

First published 2010 by
PALGRAVE MACMILLAN

Palgrave Macmillan in the UK is an imprint of Macmillan Publishers Limited,
registered in England, company number 785998, of Houndmills, Basingstoke,
Hampshire RG21 6XS.

Palgrave Macmillan in the US is a division of St Martin's Press LLC,
175 Fifth Avenue, New York, NY 10010.

Palgrave Macmillan is the global academic imprint of the above companies
and has companies and representatives throughout the world.

Palgrave® and Macmillan® are registered trademarks in the United States,
the United Kingdom, Europe and other countries.

ISBN-13: 978–0–230–22220–5 hardback

This book is printed on paper suitable for recycling and made from fully
managed and sustained forest sources. Logging, pulping and manufacturing
processes are expected to conform to the environmental regulations of the
country of origin.

A catalogue record for this book is available from the British Library.

A catalog record for this book is available from the Library of Congress.

10   9   8   7   6   5   4   3   2   1
19   18   17   16   15   14   13   12   11   10

Printed and bound in Great Britain by
CPI Antony Rowe, Chippenham and Eastbourne

For Lucy Grace

Contents

Acknowledgments

This collection of essays has it roots in a series of papers presented at the Arizona Center for Medieval and Renaissance Studies Conference in February 2007. Some of the contributors, Derek Neal, Katherine Smith, Andrew Miller, Janelle Werner and Tony Perron, delivered presentations which were the genesis of the work presented in this volume. Along the way, Scott Wells, Andy Romig, Andrew Holt and Tanya Stabler Miller joined the group, contributing their ideas on the medieval clergy and masculinity. Overall, they are an outstanding group, and I sincerely appreciate their cooperation, innovation and professionalism on this project. I say, with great pride, that these scholars have created a body of historiography that will undoubtedly shape the field for years to come.

I would like to thank Ruth Karras, Jacqueline Murray and Patricia Cullum for their inspiration and for their pioneering work in the field of medieval gender studies.

The editorial staff at Palgrave Macmillan (UK) was extremely courteous and helpful in providing assistance on this project, particularly Michael Strang and Ruth Ireland. Series editor John Arnold was very supportive of this project from the very beginning, and his assistance got the volume off the ground and into production. I also wish to thank the anonymous readers, whose feedback strengthened the volume as a whole.

The Institute for Research in the Humanities at UW-Madison awarded me a fellowship, supported by the College of Letters and Sciences and Department of History at UW-Whitewater, during which time I was able to make progress on the creation of this volume. I also wish to thank Houghton Library at Harvard University for allowing me the use of the Calderini Pontifical (MS TYP 1) for the cover image.

Virginia Galloway assisted me with proofreading and editing. Finally, my thanks to Ben for his support and advice on this project from the beginning. I've dedicated this work to Lucy, so that one day she will understand why her mother is so fascinated with medieval priests.

List of Abbreviations

MGH	*Monumenta Germaniae Historica* (Hannover: 1828–)
DD CI	Diplomata regum et imperatorum Germaniae: Conrad I diplomata
DD HI	Diplomata regum et imperatorum Germaniae: Heinrici I diplomata
DD OI	Diplomata regum et imperatorum Germaniae: Otto I diplomata
SS	Scriptores
SrM	Scriptores rerum Merovingicarum
SrG	Scriptores rerum Germanicarum in usum scholarum
PL	*Patrologia Latina*, edited by J-P Migne, 221 volumes (Paris:1844–64)
RHC Occ	*Recueil des Historiens des Croisades*, 16 volumes in 17 (Paris:1841–1906), *Historiens occidentaux*

Notes on Contributors

Andrew Holt is a PhD candidate in the History Department at the University of Florida, with a focus on religion, violence and gender in the crusades. Andrew is the co-editor (with James Muldoon) of *Competing Voices from the Crusades* (2008) and the author of several articles on the crusades. He is currently working on a project tentatively entitled *Seven Myths of the Crusades*, which explores the divergence of popular and scholarly perceptions of the crusades.

Andrew G. Miller (PhD, University of California-Santa Barbara) is a visiting assistant professor at DePaul University in Chicago, Illinois. His current area of research is focused on the uses and significance of violence and violent discourses in the construction of clerical and lay masculinities and identities in medieval Europe, especially England.

Tanya Stabler Miller (PhD, University of California-Santa Barbara) is Assistant Professor of History at Purdue University-Calumet. Her research interests focus on issues of male clerical identity, female religious culture and urban piety in medieval Paris. She is currently engaged in a study of beguine communities in medieval Paris and their relationship to the city's scholars, merchants and nobles. She is the author of 'What's in a Name? Clerical Representations of Parisian Beguines (1200–1328)', *The Journal of Medieval History* 33 (2007).

Derek Neal (PhD, McGill University) is Assistant Professor of History at Nipissing University in Ontario, Canada. His book, *The Masculine Self in Late Medieval England*, was published in 2008. Derek was recently awarded a Social Sciences and Humanities Research Council of Canada Standard Research Grant for a new project investigating the gendered quality of lay–clerical relations in England c. 1460–1560.

Anthony Perron (PhD, University of Chicago) is Associate Professor of Medieval History at Loyola Marymount University in Los Angeles. He has published several articles on the church history of high-medieval Scandinavia and is the author of a chapter on the papacy in volume four of the Cambridge History of Christianity.

Andrew Romig (PhD, Brown University) is Lecturer and Assistant Director of Studies for the Committee on Degrees in History and Literature at Harvard University. His work focuses on Carolingian lay religiosity and the intersections of secular and monastic culture during the eighth, ninth and tenth centuries. He is currently at work on a monograph exploring the cultural trauma of the Carolingian civil war (840–843) and a translation of the *Opus Caroli regis contra synodum*.

Katherine Allen Smith (PhD, New York University) is Assistant Professor of History at the University of Puget Sound. Her articles on monastic devotional and penitential practices have appeared in *Speculum*, *The Journal of Medieval History*, *Viator* and *Church History*. She is currently at work on a monograph which explores representations of war in monastic texts from c. 950 to c. 1200.

Jennifer D. Thibodeaux (PhD, University of Kansas) is Associate Professor of History at the University of Wisconsin-Whitewater. She specializes in the study of gender identity among the medieval Norman clergy and has published articles in *Gender and History* and *Essays in Medieval Studies*. She is currently writing a monograph tentatively entitled *The Manly Priest: Masculinity and the Clergy in Medieval Normandy, 1050–1300*.

Scott Wells (PhD, New York University) is an associate professor of History and Director of Religious Studies at California State University, Los Angeles. With Katherine Allen Smith, he has co-edited a collection of essays in honor of Penelope D. Johnson entitled *Negotiating Community and Difference in Medieval Europe: Gender, Power, Patronage and the Authority of Religion in Latin Christendom* (2009). His current projects include a study of the community of canonesses at Gandersheim from the ninth through the twelfth centuries, as well as a monograph entitled *From Chronicling Monks to Prophesying Nuns: The Search for History's Design in Medieval Germany*.

Janelle Werner (PhD, University of North Carolina-Chapel Hill) is Assistant Professor of History at Kalamazoo College. She held an Andrew W. Mellon/ACLS Early Career Fellowship for 2009–2010. She is currently revising her dissertation on priests and concubines in late medieval England, directed by Judith Bennett.

Introduction: Rethinking the Medieval Clergy and Masculinity

Jennifer D. Thibodeaux

In May 2009, the Roman Catholic Church was once again hit with scandal when the Reverend Alberto Cutié, priest of Saint Francis de Sales in Miami, Florida (USA), admitted to having an affair with an unnamed women for more than 2 years. Father Cutié, described by his parishioners as a charismatic and handsome 40-year-old man, told a local newspaper in 1999 that celibacy was 'a struggle, but it's a good struggle.'[1] Ten years later, photographs published in local media outlets showed Cutié kissing and embracing a woman on a Miami beach. After a public apology and his subsequent removal from his parish church, Cutié was quoted as saying, 'I didn't stop being a man just because I put on a cassock. There are trousers under this cassock.'[2]

Cutié's remarks before and after the scandal perfectly exemplify the tension implicit in clerical masculinity, both for the Middle Ages and for our own time. Masculinities are subject to definition and negotiation in every society, culture and time period. Cutié's statements express the conflict inherent today between being a 'man of the Church', a masculinity dependent upon a medieval institution, and being a man of twenty-first century America, a masculinity inextricably tied to all things modern. Such genders appear as nicely packaged, cohesive identities, and yet both are far more complex than they appear, having evolved through a series of internal and external negotiations and challenges. One thousand years ago, the clergy of the Roman Catholic Church struggled to define manliness, and this is the focus of this collection. As the Catholic Church currently faces a shortage of priests to service its parishes worldwide, the issue of celibacy and marriage for the priesthood continually resurfaces as a pertinent and controversial question. Inevitably, the perceived link between sexuality and masculinity affects the way in which scholars have viewed the gender of the

Catholic clergy, both medieval and modern. Contemporary studies of masculinity emphasize that there are multiple standards of the masculine gender, and how they are defined varies according to culture.[3] Sexuality, as crucial as it sometimes appears for the performance of masculinity, is not the sole marker of manhood. As Cutié alluded to by his comment in 1999, the Catholic Church views clerical celibacy as a masculine accomplishment, a signifier of virile strength in the face of carnal weakness. This is exactly the way in which the medieval Church conceived of celibacy. But while celibacy and sexual performance continue to be pertinent components of masculinity, masculine gender is performed in numerous ways. Studying the medieval clergy forces the historian to look outside sexuality for how men defined their masculinity. In this way, the theoretical framework employed in examining medieval clerical masculinity can be extended well beyond the Middle Ages. It allows scholars to break open the category of 'man' and question how the masculine gender is manifested, challenged and negotiated in any given time or historical period, by any group of men. Thus, our studies of clerical masculine performances illustrate the varieties of masculinities and in turn prevents the 'universalizing' of the male experience.[4]

The gender identity of the clergy makes for a fascinating and perplexing historical problem, one not yet addressed sufficiently in the scholarship of the field. Ruth Karras, in the first monograph on medieval masculinity, demonstrated how young men, such as knights, journeymen and university students, learned masculine performance according to their social status and occupation.[5] Examining clerical gender requires a different set of approaches, because clerical rank, status and occupation varied considerably as did the requirements of the vocation. While the essays presented here are not exhaustive in scope, they do begin the examination of how clerical masculinity was defined in different time periods and geographical context. Was a cloistered monk considered male in the same way as a parish priest? How did a scholastic theologian perform masculinity while competing with Franciscans and Dominicans for students in thirteenth-century Paris? How did eleventh-century German monks respond to the militaristic masculinization of their order? For all clerics, did their self-identification as masculine conflict with how their communities viewed them? How did these men negotiate their masculine identities, while abiding by celibacy and other prohibitions on manly behavior? The essays in this collection assume that not only did the medieval clergy construct masculine (not feminine) identities, but they did so in different ways, whether through sexuality, celibacy, violence or other gendered behavior. Another nuance of studying clerical gender is that not only did clerics express their masculinities in different ways, but their respective notions of manliness often conflicted. Clerical identity was a gender of multiplicity, and one vigorously negotiated and contested by the Church, its clergy and medieval society.

Medieval historians have not ignored the question of masculinity. In 1994, Clare Lees' collection, *Medieval Masculinities: Regarding Men in the Middle Ages*, brought together scholarship from different disciplines to consider the question of how medieval males performed masculinity. Included in Lees' collection was the highly influential article by the late Jo Ann McNamara, 'The *Herrenfrage*: The Restructuring of the Gender System, 1050–1150', which placed at the forefront of this problem the question of medieval clerical gender identity in the context of church reform.[6] After *Medieval Masculinities*, other volumes examining medieval masculine gender soon emerged, and most included essays on clerical gender identity: *Becoming Male in the Middle Ages*, edited by Jeffrey Cohen and Bonnie Wheeler (1997); *Masculinity in Medieval Europe*,[7] edited by D.M. Hadley (1999); and *Conflicted Identities and Multiple Masculinities: Men in the Medieval West*,[8] edited by Jacqueline Murray (1999). While these collections did include some notable essays on clerical gender identity, scholarly work in this area has lagged behind other studies of medieval gender, especially in the discipline of history. More recently, *Holiness and Masculinity in the Middle Ages*,[9] edited by P.H. Cullum and Katherine J. Lewis (2005), has come closest to approaching the important questions on clerics and masculinity, yet overall focuses on holiness and sanctity as the basis for gender performances.

In the absence of monographs on clerical masculinity, the sparse historiography of the field has been built largely from key essays in these collections. In her monumental essay, JoAnn McNamara raised an important issue which lends itself precisely to the debate on clerical gender identity: 'Can one be a man without deploying the most obvious biological attributes of manhood?'[10] McNamara's key thesis reflected the trajectory of initial scholarship on clerical masculinity, which underscored the centrality of celibacy as the most vital component of a cleric's gender identity. After all, if one employs Vern Bullough's universal tripartite model of manliness (borrowed from anthropologist David D. Gilmore), one in which men are defined as 'impregnating women, protecting dependents, and serving as provider to one's family', a celibate group of men, like the medieval clergy, will immediately fail the test of manhood.[11] In several articles from these collections, scholars have explored the gendered lives of clerics, and, with few exceptions, viewed them by the standard of sexuality, and suggested that clerics represented a feminine or ungendered group. Celibacy itself has been the principal category of analysis. Many scholars assume that when the Church prohibited clergy from pursuing sexual encounters, these men were crippled in their attempts to acquire a masculine identity. McNamara argued in 1994 that the enactment of celibacy upon the clergy created a 'herrenfrage' (masculine crisis), as both celibate and non-celibate men struggled to redefine masculinity.[12] R.N. Swanson, in one of the most popularly cited articles in the field of medieval masculinity, has described the gendered state of clergy as one of 'emasculinity', an in-between stage of gender identity between masculinity and femininity, or as anthropologists see it, a 'third gender'.[13]

Viewed broadly, these assumptions about the medieval clergy appear plausible. Yet, the most important contribution of these essays lies in the historiographical foundation they created which has led scholars to continue this discussion, building upon these analyses and extending or modifying these arguments as new historical work in the field emerges. The more recent scholarship on clerics and masculinity has provided a much different framework for understanding the field. Scholars, such as P.H. Cullum and I, adapting the work of modern gender theorist R.W. Connell, have argued that the gender of medieval clerics is best viewed as one of multiplicity, rather than a monolithic standard. In our respective works, we have shown how parish clerics in England and Normandy, far from seeing themselves as emasculated or part of a third category of gender, instead identified with laymen in their communities and adopted those standards of masculinity.[14] As Cullum has noted, the secular clergy suffered from a 'more fragile gender identity' than that of monastic clergy, due to the relatively late acquisition of clerical masculinity and the existence of more alternatives. Even monks in the late Middle Ages may have suffered some identity crisis as they were entering monastic communities well after the formative period of gender acquisition.[15] In my own work on the Norman clergy and gender, I have shown how the efforts of reform by the Archbishop Odo Rigaldus failed to 'monasticize' the parish clergy due to an incompatibility between the masculine values held by the men of the village and the ideals of clerical manhood enforced by the reforming Church. With the large majority of parish priests drawn from their local communities, a conflict of gender identity occurred, forcing these priests to choose between being a 'man of the Church' or a 'man of the village'.[16]

What exactly was the state of the conflict? What made the medieval clergy as a group of men so different from their lay counterparts? The short answer is, of course, celibacy. Celibacy, however, was but only one component of an institutional idea that I have called 'clerical manhood'.[17] Beginning with the so-called Gregorian reform movement of the eleventh and twelfth centuries, ecclesiastical reformers sought to elevate the clergy in status and separate them from the laity. Part of this effort required a 'monasticization' of the priesthood by the imposition of celibacy, a requirement only previously applied to the monastic orders. Celibacy was important for two reasons. First, as the sacramental nature of the Eucharist developed, reformers saw the sexuality of priests as a corrupting, even at times a polluting, presence at the altar.[18] Secondly, parish priests, and even some members of the upper clergy, historically had passed their church benefices to their sons, occasionally even providing the same as dowries to their daughters upon marriage.[19] The Gregorians aimed to prevent such alienation of church property through hereditary dynasties; this was, in fact, a primary goal of the reform movement, which sought to remove lay control of church offices permanently. The celibacy requirement thus functioned on both a theological and economical level for the Church.

While the Gregorians had valid reasons for enforcing celibacy upon the secular clergy, the imposition of such a requirement did, at times, meet with violent resistance. The secular clergy, those ordained to the priesthood as well as other clerics, operated in the world, undertaking the 'cure of souls' (*cura animarum*) and pastoral ministry. Rather than being separated away from the laity, they lived among their parishioners in communities that emphasized traditional notions of masculine and feminine gender. Thus, at a time when their peers were marrying and acquiring households, the parish priest was supposed to be a 'man of the Church' instead, to live a celibate and separate life from his lay counterparts. Medieval monks traditionally were cloistered away in their religious communities, generally limited in their interactions with local, outside village communities. In the early Middle Ages, the tradition of child oblation meant that many monks developed their gender identities inside the cloister. Enforcement of clerical manhood should have been easier among the monastic clergy, yet many gendered conflicts arose within this vocation as well, especially as the era of child oblation died away, and men joined monastic communities as adults, carrying with them their ingrained notions of secular masculinity.

Celibacy, however, was not the only criterion for clerical manhood. Ecclesiastical reformers created and with varied success tried to enforce a clerical model of manhood, one that could apply to all members of the clergy, from the cloistered monk to the parish cleric. Recognizing the compelling roles of medieval men as husbands, fathers and warriors, church authorities created spiritual equivalents for their clergy, encouraging them to be husbands, fathers and warriors *of the Church*. Thus Gregorian reformers wanted more than the simple imposition of celibacy. Ecclesiastical reformers promoted a vision of the clerical life as one of manliness. This model, while centered on the imposition of clerical celibacy, involved a code of conduct that either cut ties with or modified secular conceptions of masculinity; its goal was to elevate and differentiate the clergy in status from ordinary men, non-clerics who lived in the secular world. Thus, clerical manhood (or clerical masculinity) existed as an ideal, promoted by reformers who wanted to purify the ranks of the Church by designating a code of behavior that emphasized clerical spiritual authority. They created this ideal, in part, on the denial of socially recognized forms of masculine behavior. In some cases, however, they incorporated traditional masculine values into their spiritual models.

While the notion of 'clerical manhood' was predicated upon spiritual authority and power, it still remains to be examined how this ideal varied within the minds of ecclesiastical reformers. To the historian, priests and monks are the most visible and most recognized as clerics of the medieval Church; yet there were many kinds of clerics: some celibate and some non-celibate, some cloistered and some active in pastoral ministry, some members of the austere Cluniacs and some members of the traditional Benedictines. It is important to consider such nuances when examining the medieval clergy, especially from the perspective of gender. Just as we would not expect English knights to perform masculinity in the same way as French journeymen, we should not expect medieval clerics to perform masculinity identically. For instance, the manner in which cloistered Cistercian monks expressed their spirituality may have been gendered differently than that of the itinerant Franciscan friars.[20] Our conclusions about the masculinity of a parish priest in rural Normandy would not necessarily, and probably not at all, extend to German monks of the eleventh century. Similarly, we may not expect that the masculine power defined by a bishop of thirteenth-century England was expressed or conceived of in the same way as that of Danish priests from the same era. Clerical group, time period and geographical context all influenced the variable enactments of medieval clerical masculinities. The institutional definition of clerical manhood may have been a singular ideal, broadly defined by spiritual authority, but the ways in which it was negotiated and contested by different groups of clergy was wide-ranging. The various clerical subgroups (monks, priests, secular and regular canons, bishops and mendicants) modified these roles as appropriate to their vocation and in response to perceived deficiencies. Altogether, the essays in this volume elucidate the dialectical formation of clerical gender identity and show how the various groups of clerics deviated from one another in their interpretation of masculine performance.

Thus, the clerical ideal of manliness, created by a reforming Church, focused on more than celibacy; it operated in tandem with secular conceptions of manliness. Likewise, Derek Neal argues in his essay, 'What Can Historians Do with Clerical Masculinity? Lessons from Medieval Europe', historians should do more than rely on celibacy as a category of analysis for medieval clerics. Neal warns against classifying groups as 'unmasculine' based on the criteria of other groups, much in the way that R.N. Swanson did in his seminal essay. After all, if we judge a celibate monk to be unmanly because he was viewed as such by the adventuresome, violent and sexually promiscuous knight, we must remember that the reverse was also true. From the perspective of the celibate monk, the knight suffered from a lack of self-control and possibly effeminacy by giving into the lustful advances of women. Studying medieval clerical masculinity forces historians to think of gender as less a concise, delimited package of meaning and more a concept with multiple historical definitions. Neal also suggests that historians move beyond the traditional methodology employed in studying the past. The psychological aspect of individuals should not be ignored, as it is important to remember that for the medieval priest, taking Holy Orders did not suddenly 'cancel out' the self-identity constructed in childhood and adolescence, a theme I explore in my own contribution to this volume.

Was this the crisis that McNamara spoke of? Perhaps. These essays make clear celibacy was not the only factor in clerical gender identity. Nor, for that matter, was enforcing clerical manhood on all groups of clerics a simple task. It was instead a process fraught with conflict and negotiation, both within clerical performances of gender and with secular masculinities. The essays presented here continue with this theoretical perspective, by examining the gender of medieval clerics as one of multiplicity, further interrogating the notion of clerical manliness within particular contexts.

I Monastic masculinity: Bridging secular and clerical identities

The first theme explored by this volume is the attempt by church reformers to bridge the gap between the secular and clerical masculinities. This theme is most apparent when studying the monastic clergy of the early and high Middle Ages. Both before and after the period of ecclesiastical reform in the eleventh and twelfth centuries, clerical reformers and secular leaders were cognizant of the difficulties imposing a code of behavior that so rigidly conflicted with society's gendered ideals. As numerous saints' lives (*vitae*) from the tenth and eleventh centuries indicate, many aristocratic men, desiring the spiritual life, left behind their secular lives of marriage and war to enter monastic communities. These *vitae* support the monastic life as one that was superior to the worldly values of the aristocracy. Instead of a complete repudiation of secular manhood, the ideal of monastic manliness changed to accommodate the values of secular masculinity, and quite possibly, to make easier the transition from the outside world to the cloister. At the same time, monastic writers elevated this concept of manliness to a hegemonic position, an effort part of the general reassertion of clerical superiority characteristic of the reform period. Due to their own goals, reformers created a model to reconcile these ideals of manliness. This negotiation of gender created new models for monastic and secular life.

Along these lines, in 'The Common Bond of Aristocratic Masculinity: Monks, Secular Men and St. Gerald of Aurillac', Andrew Romig argues that Christian manliness, as developed by the tenth-century writer Odo of Cluny in his *Life of St. Gerald*, espoused the call to public service and good works, aspects of early medieval manhood that could be shared by aristocratic lords and monks alike. Precisely in this way, Count Gerald of Aurillac became a model of manliness for both laymen and clergy. Contrary to previous scholarly assessments of Gerald's *vita*, which have argued that Gerald's model was either inappropriate or conflicted with regard to gender identity, Romig suggests that Gerald's masculinity was a model for all aristocratic men, both secular and monastic. Gerald, as presented by Odo, then transcends the gendered categories of lay and religious, and subsequently brings together both groups of men in a common bond of Christian manliness. Through his analysis, Romig becomes one of the first scholars to include the category of social status in his definition of clerical masculinity.

At other times, such a common bond of manliness, one which brought together different social groups, was not accepted by all reformers. As Scott Wells illustrates in 'The Warrior *Habitus*: Militant Masculinity and Monasticism in the Henrician Reform Movement', Henry II, German emperor of the eleventh century, sought a masculinization of the monastic order in his realm. Wells finds gender ideology in a most unusual source: the texts of grants of privileges and properties to royal monasteries by the emperor. In these documents, Henry was able to make clear that the cenobitic life was like a military unit, both hierarchical and comprised of soldiers in the service of God. The *vitae* of Henry's reforming abbots also served to underscore the emperor's masculinist agenda by emphasizing militant masculinity. Yet Henry's program met with resistance as indicated by contemporary monastic chroniclers. The monks in Henry's realm saw monastic life as one that closely held to the ideals of harmony and egalitarianism, not one that constructed the cenobitic life into a warrior *habitus*.

Despite the experience of Henry's monks, the warrior *habitus* was a concept frequently employed by monastic writers in early northern France, amply shown by Katherine Smith in 'Spiritual Warriors in Citadels of Faith: Martial Rhetoric and Monastic Masculinity in the Long Twelfth Century'. In Smith's analysis, the masculinity of twelfth-century monks did not develop in a 'cloistered vacuum' but was the product of a dialogue between the secular and religious worlds, and this explains the use of military imagery in hagiography of the period. Hagiographers' use of a martial rhetoric in the *vitae* provided monks with a military genealogy within the Church and helped them define their role within the hierarchies of the medieval world. Furthermore, monks did not present themselves as virile in order to make up a perceived deficiency in their model of manliness; instead, as Smith argues, virility was a reflection of their confidence and commitment to a more dangerous battle than those fought on earth. This ideal of manliness, of manly prowess and virtue, was such a compelling and ubiquitous ideal that it extended even to female monastics.

II Priestly masculinity: Negotiating celibacy and sexuality

Recognizing that celibacy has compelling significance for clerical identity, the authors of these essays underscore the extent to which parish priests negotiated the role of sexuality in their gender performances, and the degree to which celibacy and sexuality functioned as important markers of masculine identity for these particular clerics. The experience of the secular clergy, both priests and other clerics, was quite different than that of the cloistered monk. While celibacy had always been a feature of monasticism, parish priests had a history of creating biological families, and cohabitating with wife and children. During the era of reform, clerical marriages were invalidated and clerics in major orders (subdeacon, deacon and priest) were required to live celibate lives. Clerical wives became clerical concubines, and the children of priests were rendered illegitimate by these canonical changes.[21]

From a practical perspective, the overnight abolition of clerical families was unrealistic. Furthermore, the model of priestly celibacy proved difficult to implement. Looking at a geographic area outside of Latin Christendom, Anthony Perron asserts in 'Saxo Grammaticus's Heroic Chastity: A Model of Clerical Celibacy and Masculinity in Medieval Scandinavia', that no paradigm of sacerdotal celibacy existed in Scandinavia until Saxo created an ideal of heroic chastity in his *Gesta Danorum*. The Scandinavian Church was a young institution, with few rules that governed its clergy. Like other men, Danish priests looked back to the heroes of their mythic origins for models of manliness. Saxo subtly encouraged sacerdotal celibacy by constructing Scandinavian warriors as chaste kings rather than as sexually virile men. Such figures were portrayed in his work as battling the enticements of womanly affection in order to resist being 'softened'. Saxo depicted women, both royal wives and daughters, as dangerous threats to the manliness of kings. The message of his work becomes clear: asceticism was associated with manliness while sexuality was associated with effeminacy. Saxo's model not only reinforced the Gregorian ideal for priests, but it also created a model of manliness for the Danish clergy, one which was designed to inspire the clergy to lead celibate lives while emphasizing the corrupting effects of sex.

Chastity was presented as more than a heroic ideal to the Norman priests of the thirteenth century. Ecclesiastical reformers promoted celibacy as a marker of social adulthood that clerics could claim upon their ordination to the major orders. Yet, although celibacy was preached as a marriage to the Church, many clerics did not see it as the effective, monumental marker of masculine adulthood. As I demonstrate in my essay, 'From Boys to Priests: Adolescence, Masculinity and the Parish Clergy in Medieval Normandy', some Norman clerics, unable to reconcile celibacy and social adulthood, behaved as adolescents even as they aged biologically. It has been previously argued that clerical adulthood was reached upon acquisition of a benefice, the counterpart to a household. I argue against this assumption and examine how peer and community approval factored more importantly in the acquisition of a clerical identity and social adulthood. For many priests, their participation in male bonding influenced their behavior and their identities far more than the institutional regulations of the Church. By examining the behavior of these clerics, it is possible to find glimpses of the masculine psyche, especially in those who, although they presided over a parish church, refused to be ordained to the order of priesthood.

Did some parish priests intend to construct a new, clerical masculinity, one which incorporated concubinage as a part of this identity? This is the question pursued by Janelle Werner in 'Promiscuous Priests and Vicarage Children: Clerical Sexuality and Masculinity in Late Medieval England'. Werner argues for a different approach to examining celibacy within the context of clerical life, especially that of parish priests in late medieval England. Werner suggests that instead of viewing clerical manliness as an ideal at odds with secular manliness, scholars should begin looking at how clerics incorporated sexuality into their clerical identities. Some English priests so flagrantly ignored their vows of celibacy that it appears not to have been a crucial factor in their performance of clerical masculinity. Instead of repudiating lay masculinity, English priests performed their gender as if it was an identity which overlapped the clerical and lay spheres of life. Werner uses statistical evidence to show that English priests involved in sexual relationships were a sizeable minority. If English historians have equated this behavior with immorality or refused to see it as an important issue then, as Werner argues, they have neglected a very real, meaningful experience that pertained to the gender identity of these men.

III   Clerical masculinity: Contested identities

The final theme of this volume is the negotiation of gender within a context of conflict, between secular and clerical performances of manliness. Masculinities are always held in tension, as they compete with one another for hegemony. This struggle for hegemony was enacted not only between secular and clerical masculinities, but also between competing clerical masculinities. The essays in this section highlight the gendered identities which existed outside of the cloister and parish but which were still clerical. Bishops, university theologians and crusaders were such men whose vocations demanded a different kind of gender identity than those performed by cloistered monks and parish priests.

Perhaps no other group of medieval men so confound the definition of clerical masculinity and secular masculinity as the medieval crusader. As Andrew Holt demonstrates in his essay, 'Between Warrior and Priest: The Creation of a New Masculine Identity during the Crusades', the competing masculine ideals of the clergy and knightly class were individually insufficient in addressing the unique position of the crusader. On the one hand, the crusade was a spiritual endeavor, one that required the warrior to adhere to certain spiritual values. On the other hand, the crusade was a war, one that depended upon bloodshed and various acts of violence. Herein lies the difficulty in analyzing the gender of such men. Both knight and cleric, being a crusader meant employing restraint in sexual matters, yet fighting 'manfully' to protect Christendom. Clerical writers, desiring to promote the crusader as a fierce, yet humble, warrior created what Holt calls a 'hybrid' model of manliness. This hybrid model included a combination of humility, devotion and chastity, which coexisted with bravery and military skill. Utilizing masculine aspects of both groups, clerical writers elevated the gender of the crusader to embody what they viewed as the best of both masculinities. Holt breaks new ground as he applies a gendered analysis focused on masculinity to a much-studied subject and demonstrates that crusader masculinity was indeed a carefully negotiated gender identity, contested by secular and religious men alike.

In an age when bishops were 'monasticized', many prelates faced a conflict between clerical manhood and corporate leadership of their churches and communities; in a world where secular lords dominated and defined gender, medieval bishops struggled with performing masculinity. Clerical masculinities were frequently contested by secular masculinities, as numerous conflicts between religious men and powerful lords in the Middle Ages indicate. If secular masculinity was understood by the trinity of marriage, procreation and providing for one's family, then as Andrew Miller argues in 'Knights, Bishops and Deer Parks: Episcopal Identity, Emasculation and Clerical Space in Medieval England', episcopal manliness might be understood as marriage to the Church, protecting household dependents of the episcopal manor, and providing for one's diocese and household. In thirteenth-century England, however, bishops faced numerous challenges from powerful laymen opposed to this episcopal standard of masculinity. Miller argues that English knights' violent attacks on episcopal deer parks was not only a method of repudiating episcopal power, but also a gendered ritual of emasculation designed to highlight the impotence of the bishop as head of his household (and diocese). The response of bishops to these attacks was nothing less than paternal, as these prelates used the weapon of excommunication to assert their manly authority while simultaneously defeating the power (and hegemony) of the knights.

Masculine competition existed not only between secular and clerical masculinities, but also between clerical groups, as Tanya Stabler Miller demonstrates in 'Mirror of the Scholarly (Masculine) Soul: Scholastics, Béguines and Gendered Spirituality in Medieval Paris'. Stabler Miller takes an innovative approach to the study of university scholastics by analyzing their gender in the context of their relationships with the beguines, laywomen who led religious lives. The form of clerical masculinity embraced by scholastic theologians conflicted sharply with the clerical responsibilities of pastoral care. This conflict became crucial at a time when secular clergy faced vocational competition from the mendicant preachers. Stabler Miller uses the case of Robert of Sorbon, a thirteenth-century secular cleric, whose writings sought to negotiate this divide between scholastic theologians and mendicant preachers by utilizing aspects of the beguine's spirituality. Rather than associate with beguines, and risk masculine status, Robert adopted aspects of their religiosity to 'soften the aggressive masculinity' of the secular clerics. Robert preserved the clergy's pastoral authority by deploying the masculine term *béguin*. As Stabler Miller shows, Robert was not successful in implementing this new model of clerical masculinity. His students instead preferred to represent the beguine as a separate, feminine model to contrast with the clerical (masculine) emphasis on reason. Ultimately, the cleric did not become a *béguin*, but still used the *béguine* as a spiritual model.

All of these essays show how varied and complicated clerical gender identity actually was in the Middle Ages. This collection goes farther than previous assumptions that the medieval clergy was ungendered, feminine or something else. Instead, it builds upon gender historians' recent insight that masculinity is not singular, but a gender of multiple definitions, constantly in tension and always subjected to negotiation. To define the clergy simply from one masculinity alone creates a false historical construct because the clergy was not itself a singular entity. The value of this line of thinking extends well beyond the Middle Ages. Studying the medieval clergy shows scholars in other fields how to radically deconstruct masculine gender. In particular, investigating clerical gender underscores how to bypass sexuality as the sole constitutive element of masculinity, or perhaps even as its dominant element. The questions raised by the essays in this volume easily extend to studies of the clergy in other time periods, and more broadly, to the study of non-clerical men. Altogether, this volume lays a solid historiographical foundation for more detailed studies on medieval clergy and masculinity in the future.

Notes

1. Tim Padgett, 'The Father Cutié Scandal: Sex and the Single Priest', in *Time Magazine*, 7 May 2009, http://www.time.com/time/nation/article/0,8599,1896581,00.html, accessed online 22 September 2009.
2. Reuters, 'Sex scandal Miami priest quits Catholic Church', http://www.reuters.com/article/oddlyEnoughNews/idUSTRE54S4FT20090529, accessed online 28 May 2009.
3. R.W. Connell, *Masculinities* (Berkeley, CA, 1987) and David D. Gilmore, *Manhood in the Making: Cultural Concepts of Masculinity* (New Haven, CT, 1990).
4. See the essay by Jacqueline Murray, 'Hiding Behind the Universal Man: Male Sexuality in the Middle Ages', in *Handbook of Medieval Sexuality*, edited by Vern Bullough and James Brundage (NY, 1996), 123–52.
5. Ruth Mazo Karras, *From Boys to Men: Formations of Masculinity in Late Medieval Europe* (Philadelphia, 2003).
6. JoAnn McNamara, 'The *Herrenfrage*: The Restructuring of the Gender System, 1050–1150', in *Medieval Masculinities*, edited by Clare A. Lees (Minneapolis, 1994), 3–29.
7. See the essays by Conrad Leyser, 'Masculinity in Flux: Nocturnal Emission and the Limits of Celibacy in the Early Middle Ages', 103–20, and Janet Nelson, 'Monks, Secular Men and Masculinity, c. 900', 121–42, in *Masculinity in Medieval Europe*, edited by D.M. Hadley (London and NY, 1999). The essay by P.H. Cullum and R.N. Swanson will be further discussed below.
8. See the essays by Megan McLaughlin, 'Secular and Spiritual Fatherhood in the Eleventh Century', 25–43, and Jacqueline Murray, 'Mystical Castration: Some Reflections on Peter Abelard, Hugh of Lincoln and Sexual Control', 73–91, in *Conflicted Identities and Multiple Masculinities: Men in the Medieval West*, edited by Jacqueline Murray (NY, 1999).
9. In particular, see Jacuqeline Murray, 'Masculinizing Religious Life: Sexual Prowess, the Battle for Chastity and Monastic Identity', 24–42, and Robert Mills, 'The Signification of the Tonsure', 109–26, in *Holiness and Masculinity in the Middle Ages*, edited by P.H. Cullum and Katherine J. Lewis (Aberystwyth, 2004). This volume contains many other essays concerning masculinity and religious life.
10. McNamara, 'The *Herrenfrage*: The Restructuring of the Gender System, 1050–1150', in *Medieval Masculinities*, 5.
11. Vern Bullough, 'On Being Male in the Middle Ages', in *Medieval Masculinities*, 34; David D. Gilmore, *Manhood in the Making*, 223.
12. McNamara, 'The *Herrenfrage*', *passim*.
13. R.N. Swanson, 'Angels Incarnate: Clergy and Masculinity from Gregorian Reform to Reformation', in *Masculinity in Medieval Europe*, edited by D.M. Hadley (New York, 1999), 160–77.

14. P.H. Cullum, 'Clergy, Masculinity and Transgression in Late Medieval England', in *Masculinity in Medieval Europe*, 178–96; Jennifer D. Thibodeaux, 'Man of the Church or Man of the Village: Gender and the Parish Clergy in Medieval Normandy', *Gender and History* 18/2 (August 2006): 380–99.

15. Cullum, 'Clergy, Masculinity and Transgression', in *Masculinity in Medieval Europe*, 179–80.

16. Thibodeaux, 'Man of the Church', *passim*.

17. Thibodeaux, 'Man of the Church', 382–6.

18. Dyan Elliott, *Fallen Bodies: Pollution, Sexuality and Demonology in the Middle Ages* (Philadelphia, 1999), chapter four. Elliott argues that the pollution of the altar actually originates with the priest's wife, according to clerical writers of the reform period.

19. Kathryn Ann Taglia, ' "On Account of Scandal": Priests, their Children and the Ecclesiastical Demand for Celibacy', *Florilegium* 14 (1995): 57–70.

20. See the essay by Caroline Walker Bynum, 'Jesus as Mother and Abbot as Mother: Some Themes in Twelfth-Century Cistercian Writing', in her *Jesus as Mother: Studies in the Spirituality of the High Middle Ages* (Berkeley, CA, 1982), 110–69.

21. See James Brundage, *Law, Sex and Christian Society in Medieval Europe* (Chicago, 1987), *passim*, for an overview of the development of canon law on this subject.

1
What Can Historians Do with Clerical Masculinity? Lessons from Medieval Europe

Derek Neal

The title of this essay is a conscious homage to John Tosh's ground-breaking article of 1994, 'What Should Historians Do with Masculinity?', which set out the methodological issues and political concerns involved in what was then quite a new subject of historical study.[1] Tosh's eloquent argument that historians must recognize the complexity and multivalence of masculine identity has lost none of its relevance. It was, however, framed in comparatively modern terms, since Tosh, a historian of Victorian Britain, drew on nineteenth-century evidence and examples. Since the mid-1990s, medievalists and early modernists have been responding to the challenges Tosh laid out in that article. An important, if sometimes discouraging, lesson we have learned is the degree of difference in approach and assumptions between those who study masculinity in premodern history and their modernist colleagues. At some point between 1500 and 1700, as we define our periods of study, a dividing line still appears.

Clerical masculinity, the subject of this volume, has the power to revitalize connections between modern and premodern histories. On the one hand, clerical masculinity has meaning for the historical past, because as the latest research is establishing, clerics were not sheltered from contemporary gender roles and expectations. It could hardly be otherwise when their status as clerics depended on their being male. On the other hand, clerical masculinity as a category of analysis also has meaning for the historiographical present. In order to understand and use it properly, we have to push beyond accustomed definitions and habits of mind, in our considerations of both gender and history. It forces us to conceive of gender categories and historical categories less narrowly and neatly, and so has the potential to help shape a new understanding of the European past. In fact, clerical masculinity could be just the category of analysis that we need to move the history of masculinity closer to the center of gender history, and in turn to move gender history closer to the center of the historiographical discipline.

To perceive the obstacles to understanding that 'clerical masculinity' is not an oxymoron, however, we need to investigate the origins of our present perceptions. These are difficult to discount or circumvent, because they arise from attitudes about both gender and religion that have a long historical lineage. In this essay, therefore, I discuss stereotypes and preconceptions at some length, because it is not particularly easy for historians to separate themselves from the popular attitudes that suffused the eras and places they study. Instead, those attitudes become conflated with the anxieties and concerns of the historian's own cultural moment, which may well be the reason the historian wants to study masculinity at all. Shifting to a more optimistic mode, I then explore the potentialities the study of clerical masculinity has to offer.

I A contradiction in terms?

On the face of it, there could hardly be a *less* promising term for a category of historical analysis than 'clerical masculinity'. The phrase seems to short-circuit itself. A satisfactory working definition of masculinity may elude us, but the word 'masculinity' does move through our discourse trailed by a pungent aroma of associations. Even the negative ones (violence, aggression, egocentric individualism, etc.) seem to connote an instrumentality, a possibility of doing or achieving something, of being a real actor with agency in the social field. 'Masculinity', as a term, may be abstract and vague, but it is far from bland or colorless. In fact, that very capaciousness of meaning gives it an additional intrinsic potential (from *potens*, Latin for powerful, the opposite of 'impotent'). 'Masculinity' seems to describe a position where someone might get what he wants (or at the very least, gets to want something), and thereby also seems to open up a space for historical inquiry. In that space we might find people (probably men) doing things, or at least trying to do them, in a way that leaves clear and vivid imprints in the historical record. Such a history itself also opens out: it promises to yield something larger, to address broader concerns and to be important.

But when we attach, to such a promising noun, that modifier 'clerical', immediately the heart sinks, the brain dulls. 'Clerical', after all, is something akin to 'secretarial', with its overtones of the 'secret'. It evokes airless rooms smelling of paper and ink, a world of fussy rules and mind-numbing routine, of essential but dead-end, low-paying and undervalued work. A 'clerk', we think, is little more than a servant, someone who stands at a shop counter or hunches over a desk all day, jumping to do the master's bidding. A clerk inhabits a world of petty grievance, pedantry, subordination and deference; above all a world characterized, indeed defined, by *limitation* rather than possibility. The products of this 'clerical' world might be accessible to historians (since, after all, clerical people did bother to write things down and file them away, with variable neatness). But they would yield a history of interest to sub-specialists and no one else, a kind of narrow antiquarian irrelevance. And what can any of it have to do with masculinity?

Medievalists might protest that these pejorative connotations of the 'clerical' and 'secretarial' are recent – that they date (unsurprisingly) from a time, not much more than 100 years ago, when work dealing with preparing and organizing written materials (especially those conveying the desires of male superiors, and on their orders) became, increasingly and for the first time, the work of women (also, of course, the age of the 'schoolmarm' as opposed to the 'schoolmaster'). Their voices trembling with indignation, medievalists might point out that the *clerici* of the Middle Ages, at least the more fortunate of them, were the nerve center of government, politics and learning. They might even assert, with a self-satisfied flourish, that if not for all the scribbling and scratching of those 'clerks' in courtrooms, council chambers and scriptoria, we would know precious little about the deeds of kings, knights or peasants. And, they might add as the *coup de grace*, those clerks were all, every man Jack of them, a man.

The cynic, however, is unmoved. That is all very well for the Middle Ages, a world we have lost (and good riddance) but who cares, apart from the medievalists? It is not only that the Middle Ages happened so long ago, but also that they seem in some inescapable way cut off from the present day and therefore from present-day concerns such as the definitions of gender. After all, what is 'medieval' cannot, by definition, be modern, even 'early' modern; 'medieval' – 'of-the-Middle-Ages' – is a term that makes sense only in opposition to the two ends that bracket the middle. The study of gender (and particularly of masculinity), on the other hand, is inescapably modern. It depends on modern categories derived from modern evidence, and within the historical discipline, is dominated by studies of the modern world. One could argue that it makes little difference whether or not 'clerical' men more than 500 years ago possessed, thought about, problematized or rejected masculinity. Whatever their relation to the concept, it belonged to a time that is now sealed behind museum cases, in a way that even the sixteenth or seventeenth centuries seem not to be. Indeed, if 'clerical', the adjective, points toward depressing modern associations, then 'cleric', the noun, points with equivalent discouragement to those aspects of the 'medieval' world most repellent to Renaissance, Enlightenment and modernist minds (and most persistent in those of our undergraduates): Gothic fantasies, *Da Vinci Code* visions of candle-lit passages, whispered Latin phrases, an inscrutable superstructure of obscurantist knowledge in the service of a suspicious, perhaps malign, universalist Church. What 'masculinity' meant to the men who made that system work seems like a mystery better left alone.

What if we then change our terminology, speaking not of 'clerics' or 'clerical' but of 'clergy' or 'clergymen'? We then make it clear that we are referring specifically to those men who had taken Holy Orders, who were ministers of the Church in some form. Alas, the relevance to masculinity does not consequently become any clearer. The basic meaning of the word 'clergy' has not really changed over centuries, and the clergy themselves have consistently tried to speak to their own time. The apparent conflict between the religious modifier and the central concept of masculinity arises from a perception that the priorities of recognizable masculinity did not operate in the religious sphere. Because the church no longer forms the center of modern life in much of the developed world, a modern sensibility assigns its staff similarly to the margins. However able Christian ministers may be in their own sphere, that sphere seems ultimately not to overlap with the sphere in which things that matter happen (however much those in the latter sphere, politicians and judges and media personalities, may be influenced by religious convictions). So clergy and clerics in the past, we are tempted to think, lived in a culture of their own that operated according to different rules, something that made them a breed apart from the majority of laymen. For a medievalist this shows that the subject of clerical masculinity is simply subject to many of the same prejudices and stereotypes that cling to medieval history more generally. In addition, there are obstacles to its proper consideration among historians of masculinity. The period of time when clerical men might have enjoyed something like 'masculinity' is precisely the period we are least inclined to look to for provocative challenges to modern definitions.

II   No sex, no problem

In seeking the origins of these preconceptions about clerical masculinity, we may as well deal first with the subject on everyone's mind: sex, or the lack of it. Clerical celibacy was ostensibly the most obvious difference between laymen and clergymen, where it was the universal rule, and between Catholic and Protestant clergymen, where it was not. The Protestant vision of history was remarkably durable in the Anglo-American world, which was also where gender history was invented. And Protestants (including Protestant historians) are the heirs of a tradition whose founding figures hammered out their ideas of religious reform on the ground of clerical marriage. For Luther and his followers, clerical celibacy could not remain the ground of the clergy's special and privileged status, because of the lack of adequate Biblical warrant. Instead, clerical celibacy was an aberrant human invention that subverted God's plan for the Christian family. The frequency with which clergy broke their vows of celibacy, the reformers claimed, showed that the custom was contrary to the natural order of Creation, impossible for sinful humans to maintain. The ministers of the Church, rather, were called to an exemplary status as fathers and heads of Christian households. At stake was not only the behavior of individuals, but also the very constitution of church authority.[2]

For Protestant historians of subsequent centuries, therefore, it was easy to see this impossible and unnatural standard of clerical celibacy, and its violation, as one of the fatal flaws in the supposedly tottering late medieval church. That caricature, by extension, indicted the pre-Reformation clergy (and their successors in the Catholic Church) as corrupt, inclined to all manner of sexual vice under a fraudulent cover of celibacy. It did not matter that the Roman Catholic Church had attempted to address abuses in the Council of Trent; clerical celibacy marked Catholicism in Protestant eyes as mysteriously perverse, combining sexual deviance and secrecy to animate a long-lived anti-Catholic stereotype.[3]

In the nineteenth and twentieth centuries these prejudices recombined with a newly articulated discourse of natural heterosexuality. To aim for lifelong abstinence from sexual relations indicated, at best, a deficiency of sexual drive and, at worst, a sexual drive that was either repressed (to surface in some psychopathology) or perverse in another sense, directed at objects inappropriate by virtue either of their sex or age. The widely publicized exposure of sexual abuse by Catholic priests in North America, involving children of both sexes, in the late twentieth and early twenty-first centuries, has only seemed to confirm these suspicions in the minds of many people.[4]

Yet one need not subscribe to the Protestant vision of history, or to Protestantism itself, to presuppose that masculinity is not a pertinent concern for the history of the clergy (and that the clergy do not illuminate the history of masculinity). In the first place one could point to the nature of anticlericalism in Catholic countries, an anticlericalism that is highly suspicious of the sexuality of the clergy. The more the cultural standard of masculinity depends on aggressive sexuality (and the more culturally sensitive is female virginity and chastity), as in the traditional Catholic Mediterranean, the more rigid a division seems to exist between laymen and clergy. Cultural anthropologists tell us that in such countries, laymen tend not to trust priests. These are cultures where, with very specific and public ritual exceptions, piety is a largely female endeavor, both religious services and church-centered social life being dominated by women, so that the clergy are the only men outside the home with whom women regularly associate.[5]

So the present-day stereotype about clergymen's problematic sexuality, as regards masculinity, can be summed up as follows. Either clergymen are not sexually driven enough, or at all, and therefore do not qualify to be considered truly masculine. Or they are indeed sexually driven, perhaps even too much, but to the wrong objects (men or children), and therefore do not qualify to be considered truly masculine. Or they are sexually driven to the right objects (adult women), and therefore may be masculine – but do not qualify to be considered true clergymen. These assumptions wash backward in our thinking, dissolving our perception of the past into a blur of presentism.

The last possibility exposes a particularly thorny problem in historical perception and analysis: the tendency toward value judgment. Let us assume that we want to discover what masculinity meant to the clergy. Whom, then, should we study? There is, I think, a temptation to disqualify those priests and monks who broke their vows of celibacy, as insufficiently committed to their vocation. Such men were not true clergy; they were just laymen in clerical habits who should not have taken Holy Orders in the first place. (This attitude surfaced among senior students in one of my undergraduate seminars. The students were unwilling to abandon the perception of the faithfully celibate priest as the only 'real' priest.) Such judgments are very easy to make and probably help to explain why 'clerical masculinity' seems odd as a concept. This just goes to show how self-contradictory our attitudes can be: the impulse that would condemn clerical celibacy as an impossible, unnatural standard, and the one that would exclude from true clerical status all those men who did not measure up to that standard, proceed from the same point of origin: a belief that we know what the clergy are and should be, better than they themselves.

III Historians and Victorians

In fact, of course, the question of clerical masculinity involves far more than celibacy. If it did not – if the Protestant initiative to turn ministers into family men had removed the only barrier in popular perceptions – one would expect that the history of masculinity among Protestant clergy would be flourishing, and that historians would not have developed any such problematic assumptions with respect to the possibility of masculine identity among them. However, this does not seem to be the case. In fact, the Protestant clergy are conspicuous by their absence from the most recent general survey of masculinity in British history since 1500.[6]

The Reformation, one might suppose, had a double-edged and contradictory effect on the gender profile of clergymen. It gave the clergy access to all the conventional fora of masculinity (marriage, household and property) and therefore assigned to them, in even more obvious fashion than before, responsibility for all the competitive dynamics that they involved.[7] And yet, the fact that the clergy were made to live like laymen did not mean that popular perception in the long run universally accepted them as masculine. Among his Londoner father's categorical and world-view-defining suspicions, writes Brian Thompson in his recent memoir of his wartime youth, was the axiom that, plainly put, 'all vicars were poofs'.[8] One doubts that the senior Thompson was alone among his contemporaries, and I think a faint (or not so faint) echo of this belief resonates still in our gendered assumptions.

Where the Protestant clergy are concerned, certainly in the Anglo-American world, I think the crucial moment in warping our perceptions, in making us set clergymen apart, is not the sixteenth century, but the nineteenth. Here the question becomes about something larger than sex. Victorian architecture, music and manners persisted in church life until very recent years; they are there still, in some measure. (It does no good to say that such values and aesthetics represented only a privileged Anglicanism; in many cases, in America as in England, rugged Nonconformists hastened to make over their churches, their worship and their manners into that 'refined' style as soon as their congregations were rich enough to afford it.[9]) The alignment of religiosity with good bourgeois decorum was too culturally powerful to be effectively resisted; it won out in the popular mind, linking the Church and its male ministers with the soft, floral comfort and vigilant manners of the middle-class parlor, the province of the woman of the home. And, just as Victorian religion (at least, its externals) often embraced and reinforced the conformist middle-class values of Victorian society, so did the twentieth-century reaction against Victorianism – of which late twentieth-century historians are also the heirs – target a fantasized representation of the effete, stuffy and complacent Victorian clergy.

That rejection took a gendered turn, I think, even among the historians who developed the history of masculinity in the late twentieth century. Gender history, after all, was supposed to deconstruct dominant norms; it was supposed to show us how contingent gender is, how not grounded in anything natural or inevitable. Always in the background, sometimes explicitly and sometimes not, was the hopeful conviction of both feminist criticism and, further back, Marxist analysis, that the dominant norm would be shattered, overturned, by the hitherto subordinated energy of the oppressed sectors of society. By this logic the smooth, polished middle-class values that had stigmatized and excluded working-class customs as rough or uncivilized were oppressive and wrong. Clergy, by their acquiescence or enthusiastic adoption of those dominant establishment values (possibly including religion itself) were part of that hegemonic norm, and the 'muscular' type were as suspect as the non-muscular. In this vision the Victorian clergy were part of a cultural power that repressed sexuality, especially that of women, like the doctors who diagnosed hysteria and cut out women's ovaries, or those Christian reformers of prostitutes. Why give any credit to their protestations that Christianity could be manly? It was manly in the wrong way. Historians who saw themselves as in some sense rebels logically sided with the rebels, with the rude disobedient energy of the working class who rejected bourgeois religion. In fact, despite the revolutionary aim of gender studies and the constructionist insistence on gender as a deceptive fiction, this posture set up a fantasy of working-class masculinity and endorsed it as the 'right' kind.

To make things yet more complicated, though I have discussed them as erroneous stereotypes, these perceptions are not wholly deluded; there are solid historical reasons for them. For example, the impulse to place the clergy in a separate category, apart from lay society, is perfectly understandable given that for centuries on end, the rulers of the Western church attempted to accomplish exactly that. The enforcement of clerical celibacy beginning in the eleventh century, after all, came about because hard-line reformers, themselves trained in an ascetic monastic tradition that saw secular clergy as insufficiently holy, decided to impose a monastic standard on all clergy. Priests were no longer to be enmeshed in the community and family dynamics that marriage drove in secular society; instead, they were to be marked off as a special caste whose spiritual authority over the laity would be unambiguous. And so, during the High and Late Middle Ages, some sector of the higher clergy continually tried to strengthen the distinction between clergy and laity, as the institutional church grew and strengthened as a bureaucratic and judiciary entity.[10]

Similarly, the existence of a prissy stereotype in the nineteenth century did not go unnoticed by Victorian clergymen themselves. If it had, no one would have felt obliged to advocate 'muscular Christianity', and those who did were not the only ones concerned. In 1890, one Anglican priest wrote: 'I trust it is not necessary for me to say that I would have my Curate natural, free from affectation, mannerism, servility, and all those blemishes which are excrescences on manliness.'[11] One did not need to seek out worship services influenced by the Oxford Movement to encounter affectation and mannerism in nineteenth-century religious or secular culture. As for servility, it was an inescapable hazard of clerical life in any class-stratified society. This anonymous vicar, then, recognized that the clergyman often deviated from his own 'natural', plain-spoken masculine ideal in a way that set them apart from the layman. Yet in criticizing him, he effectively called for yet another kind of separateness.

Our task as historians is to be aware of all these reasons for finding clerical masculinity an odd phrase so that we can get past that feeling of strangeness. Recognizing why we classify and identify certain patterns of life the way we do, allows us to make more informed assessments of new evidence – or of old evidence, regarded in new ways.

IV   Clerical masculinity: What (not) to think about

The unhelpful aspects of these generalizations emerge because they define masculinity according to two kinds of narrow criteria that are essentially behavioral in nature. The first standard is sexual, centering masculinity on a modern heteronormative standard of male sexuality; it defines 'sexuality', too, rather narrowly in terms of sexual (which is to say, genital) *acts*. Celibate clerics, by definition, could not meet this standard.[12] More recent work has recognized the limitations of this rather simple conception, which depends not only on modern popular wisdom but on early and speculative anthropological and sociological studies as well as on a rather selective reading of historical evidence.[13]

The second standard is not totally separable from the first, although on the surface it is not a sexual definition. Its sense of masculinity is more elusive, but essentially conceives gender in terms of a degree of control and attention to the body, and of the self to the others in its world. Contrary to what one might expect, masculinity in this reading aligns with less control of the self: less control of the body's passions (including sexual ones), of the body's appearance (favoring a manly roughness as opposed to 'feminine' adornment or meticulousness) and of the body's emissions and utterances (rough language as opposed to good manners) – in all, a prizing of a kind of personal looseness rather than tightness. Again, assumptions that clerical status meant constraint to a tidy, sober primness of speech and deportment tend to disqualify clerics from this kind of masculinity. Clerical celibacy, where it obtained, would simply fit into this larger picture, restraining and damming up one among many masculine urges. Historians of lay masculinity, however, have come to realize that these definitions are too narrow, taking in only some aspects of male experience which applied only to some men at some times in their individual lives or in certain social positions. More useful and more convincing is a model of masculinity (for that matter, of femininity as well) that recognizes its multiple, distinct and yet constantly interacting dimensions: its social, physical (or embodied) and psychological dimensions, none of which is dispensable in any comprehensive historical gender analysis.[14]

Studies of the pre-Reformation clergy that address these different dimensions have already been published. Social historians have brought out the relationships between clergymen and laypeople, especially in England, that shaped clerical careers and made property and household relationships important to men in orders.[15] Other historians have explored sexual behavior among the clergy.[16] But behavior tells us only part of the story. Some historians have investigated the writings of learned clerics who thought about sexual issues and pondered the relationship between body and spirit, mind or will.[17] Still other case studies of better-documented individuals allow exploration of the psychological strata underneath these intellectual or theological writings.[18] In all cases clerics proved perfectly capable of using the discourses of masculinity, not just to criticize or preach to others, but also to speak and think about themselves.

Consequently, we have to think about masculinity as something that mattered to individuals – as an aspect of identity, in other words, not just as a set of metaphorical meanings that cultural powers could manipulate to serve rhetorical or disciplinary ends.[19] The past decade of research on masculinity has theorized and engaged the existence of multiple standards of masculinity within a given culture (hegemonic, subordinate, subaltern, etc.), any one of which 'counts' as the most important definition for the individuals exercising it.[20] In such a situation, we need to beware of classifying any individual or group as unmasculine simply because they fail to meet the standards of a completely different group. To adapt a common expression: nobody can make you unmasculine without your consent.

This issue of perspective is crucial. If we say that clerics were unmasculine, or do not count as masculine subjects, whose standards are we referring to? Take the celibate pre-Reformation clergy for example. Certainly, we can find evidence in the historical past of laymen's suspicions and resentments about clerics, manifested in casual insults and occasional violence, often in the incidental detail accompanying legal proceedings that deal with more prosaic matters around money or property. Certainly, too, there are satirical tales and poems about lustful clerics. But these bits of evidence usually do not show that laymen assumed all clerics were a different variety of male human that would function according to strange and inscrutable rules. Rather, laymen knew that clerics could, under the right circumstances, behave just like untrustworthy laymen (slandering laymen, tricking them out of their property, physically assaulting them or getting too familiar with their women), and that they were quite likely to get the better of laymen in such disputes. It seems quite plain that whatever they thought of celibacy, laymen did not use that quasi-sexual difference to exclude clerics from all of the homosocial networks that sustained a masculine social identity. Nor could they; clergymen might be their landlords or employers as well as their confessors.

Social reality also guaranteed that few clerics – apart from the most strictly cloistered monastic orders – could afford to neglect the figure they cut among laypeople, especially laymen. This situation sprang from the greatest single failure of the post-Lateran attempt to redefine the lines of Christian society. Priests indeed had to be celibate, and a judicial system for clerical scrutiny of the laity's behavior and beliefs developed. But even the most determined reforming popes and bishops never succeeded (where they tried) at the grandest level of the project: disentangling the business of becoming and being a clergyman from secular influence. A properly governed laity required good priests to minister to them, but those good men could not become priests without currying favor at some level with laypeople, they could not live without the work of laypeople and they could not be pastors if they took no part in lay affairs. So clerics needed laypeople. The other irony, of course, is that as the Church expanded and strengthened institutionally and politically, it acquired an ever more literate and educated staff whose skills were ever more valuable to secular society as well – whether to kings or to merchants. The expansion of schools and universities in Europe coincided with this period after the Gregorian reforms. So, laypeople needed clerics, and not just to hear their confessions and pray for them. In such circumstances, the failure of the effort to keep clergy apart, uncontaminated by secular connections and motives, is less surprising than the fact it was attempted at all. I have argued elsewhere that in this world, among the secular clergy especially, the kind of masculine identity that mattered was a responsibly mature one, and that celibate clerics best protected their relationships with laymen precisely by keeping their vows, indicating the degree to which laymen could trust them in everyday social relations.[21]

Marriage and sex were only part of it. Priests and monks had to take a vow of celibacy, but not all clerics did. A growing sector of literate men, clerks or *clerici* by virtue of their literacy, had neither the opportunity nor the intention of taking on actual ministry. The attempt to draw a clear line between secular and spiritual worlds had succeeded only in painting a thicker, fuzzier and grayer interstitial zone between them, a zone inhabited by men not held to the strictest clerical standard yet whose skills and livelihoods did not completely match the traditional patterns of agricultural, urban or knightly estates. Just as it does not make sense to think that masculinity meant nothing to clergymen simply because they had promised not to have sex, so it also does not make sense to think that clerks who were not clergymen should not count as 'clerical' simply because they had not promised not to have sex. We need to make sure not to be trapped by our own categories.[22]

One could still argue for a problem of masculine identity among the clergy, however. Perhaps, despite all of the mutual dependency between clergy and laity, and despite the positions of temporal advantage and spiritual authority that many clerics exercised, even honest and responsible clergymen still lived on the margins of masculine society in a different sense. Did their tonsures, their different style of dress, their adoption of a modest and sober comportment, bring on the contempt of laymen? If so, what was that contempt grounded in? Could it perhaps be a kind of jealousy originating from the fact that clerics, even those who escaped the rougher tests of youthful masculine belonging (fighting, drinking and whoring) might still enjoy masculine privileges of authority and expectation of respect – and also from the fact that laywomen might find them attractive?

That last is really more of a question about laypeople. But it does raise an important question about clerics. Did a scornful laugh from a group of young miscreants, or a suspicious look or muttered word from a pair of older laymen, strike a clergyman to the heart? Even if his position as a cleric (his connections, his reputation and his income) were not imperiled, did it still matter *to him* if his lay peers thought he was less than a man? In other words, did it make him *feel* unmanly?

Here we have reached the tough core of the question. In order even to begin assessing masculine identity in this way, we need to admit some things that historians often are not comfortable admitting. First, feelings matter: the emotional texture of life is as worthwhile to analyze historically as its tangible, quantifiable side. Second, there is a place where feelings reside, namely an individual personality animated by an individual mind. Third, that individual personality and mind are conceivable by us; we can describe them using some general terms that we more or less agree on. In other words, premodern people had the same psychological complexity that we attribute to ourselves, and that we have no problem believing existed in most humans after some mysteriously unidentifiable point in the past.

These admissions, unremarkable as they might seem, push historians into their discomfort zones. One reason, of course, is the paucity of obvious and easily interpreted evidence (especially for those who believe that quantitative analysis is the supreme form of interpretation, with traditional analytical reading a close second). Another is that it would mean adjusting a model of premodern selfhood defined almost totally in terms of external appearances, according to which, what most defined people long ago was the face they presented to others, what others thought of them, because of the potentially ruinous consequences of any loss of trust. We do not have to give up this conception completely, however, to entertain the possibility that sometimes the 'inside' could be just as significant as the 'outside'; indeed, taking account of the tension between the two would seem to offer a richer and more vivid account of historical subjectivity.

So, in order to believe that even a faithful clergyman might have a nagging sense of exclusion, we must believe that a disjuncture was possible between his clerical identity and his personal one – that having taken major or even minor orders in his early twenties did not suddenly cancel out the sense of himself as male that he had had since his earliest childhood. And, I stress again, we have to admit that there could be a 'problem' even if he did not actually have sexual relations.

Men in the stricter religious orders, in particular, struggled continually with the temptations of the world that they had left behind and yet also rarely managed to expunge from their hearts. Longing, not only for sex, but for the world's other pleasures as well (physical comfort, good food, conversation, friendly laughter, excitement and changes of scene) constantly threatened to pull them away from their path to holiness.[23] Worse yet, their bodies were not completely amenable to conscious control: erections and nightly 'pollutions' from erotic dreams might plague even the most unswerving, raising difficult questions about guilt and responsibility. As Jacqueline Murray has shown, such a battle called for strength that male monastics – arguably, in some ways, the very clerics most dramatically different from laymen – understood in explicitly masculine terms and expressed in conventionally masculine language.[24] Here, then, is a good example of the complexity of one masculine identity, integrating body, mind and society. Its male subjects experienced it as masculine, but we cannot understand it through simple categories: either of masculine versus unmasculine, or of sexual versus non-sexual concerns. And if the identity of a monastic cleric was subject to such a variety of psychological variables, surely that of a secular cleric who lived in much closer contact with laypeople must not have been any simpler. Proper consideration of these problems should prompt historians to examine what they truly believe about the composition of masculine identity, and of the respective roles of sexual and non-sexual factors within it.

V Evidence and identity

All this is not easy to do in practice. As always, there is the question of the evidence. We as historians need to work with a multidimensional concept of masculinity because otherwise we will never be able to make effective use of the evidence that tells us about it. That evidence comes from different places and appears in different forms, and often the kind we want is the kind that is most difficult to locate. Medievalists know this very well, and so are perhaps more willing than historians of later periods to make informed speculative propositions; otherwise, we would hardly ever be able to suggest anything about premodern life that was not documented in the surface detail of social reality.

We might expect better evidence for the experiences of medieval clerics than for those of their lay peers, for exactly the reasons I suggested at the outset of this essay. The fact that they were a minority of the population does not have much bearing on the matter. After all, more clerics than laypeople were literate, even if there were many poorly educated clerics, and the clergy were closer to the heart of medieval centers of record keeping. The institutional church had reason to record their existence as individuals and to keep track of what property and privileges had been granted to them, more perhaps than secular governments cared about the laity of a similar status. We might think, too, that as religious men, perhaps more of them might be inclined to reflection, contemplation and self-inquiry, which would generate arguments or propositions that we could use as clues to their interior world. Certainly, clerics generated something resembling autobiography far earlier than laymen did. Yet the historian of the clergy is subject to many of the same disappointments that pertain to lay society. There are too many records of ideals, or rules (laws, canons, regulations and instructions) and not enough records of how people actually lived within the rules, unless something went wrong and the rules were broken, prompting an institutional reaction (like a court case) – and then, we learn most about the individuals who did not live within the rules, which does not solve the problem.

Until very recently, of course, the evidence that came directly from learned clerics – their own written works, whether publications or private correspondence – was valued mostly for what it told us about the self-consciously theological, philosophical or historical concerns of its content, not for what it told us about the clerics themselves. Such evidence has the aura of importance and therefore a kind of glamour attaches to it, especially where the author is a medieval celebrity. Figures like Peter Abelard or Guibert of Nogent have been in print for a long time and their works have been extensively edited and commented upon. Reconstructing evidence about the social reality of everyday life for average clerics, however, is a slow, painstaking process, requiring the historian to risk a massive outlay of time and effort without much guarantee of a rich return. This is archive-trawling in its most potentially tedious form. It can mean weeks or months poring over documents (church court records, churchwardens' accounts, charters, bishops' registers) that often have all the zesty appeal of the average grocery list, recorded in dreadful handwriting and organized or catalogued according to no discernible method. It means attempting to track individuals by means of terse, scattered mentions of their ordinations, appointments, grants or privileges, then correlating them with whatever records may exist of their contacts with laypeople, whether in conflict (lawsuits) or cooperation (wills). Altogether, this is not a research agenda for the faint of heart, short of funding or unsure of tenure. But if medievalists can do it, anybody can.

Once that exercise has borne some fruit, we then need to work out a way of connecting it to the examination of other kinds of evidence – by which I mean especially, not only autobiographical or confessional accounts, but also the fictional literature of the period that usually gives a stereotyped image from either an idealizing or satirical extreme. Here medievalists really have something to teach the rest of the historical discipline, because they so often must rely on arguably fictional sources. This does not mean treating literature as a mimetic illustration of reality, nor does it mean using history to debunk literature. The trick is to find a credible means of drawing conclusions, not from the surface representation of a literary source, but from the buried or implicit assumptions or anxieties that have been transformed into that surface narrative or description. While this is an important project for history generally, it is particularly important for the clergy, because precisely those literate clerics who wrote about the clerical life expressed themselves in literary tropes that comprise a virtual language of their own (unsurprising given that Christianity is a religion of the book, and something that did not change with the Reformation).

These concerns about evidence and interpretation tend to apply to any attempt to conceive premodern identity for non-elite subjects. The clergy, however, add an extra challenge that springs from their status as part of both sociocultural and religious history. Our sense of clerical masculinity will not be complete unless it can take account of the religious side of being a clergyman – which might seem obvious, but it is usually not the first thing that gender historians think of. It is unsettling enough for most historians to contemplate entering the murky swamp of the psyche. How much more do we squirm, then, to be brought face to face with *piety*: a word we do not much like to use, in an age where even faithful churchgoers are not pleased to be called 'pious'.

But engaging with piety as a factor in historicized gender identity in fact helps both gender history and religious history. Piety is a matter of personal and interior belief, deeply inflected with emotion, expressing deep psychological features and therefore highly individual. It is also a matter of public observance and ritual action (indeed, even competitive demonstration), open to the scrutiny of peers, influencing and influenced by relationships with others, and therefore highly social. Whether socially or individually, piety is also undeniably gendered, making complex use of masculine and feminine imagery and models, as a great body of historical and non-historical scholarship has established.[25] Through the lens of a properly conceived piety we can see how the different dimensions of masculine identity would work together. That would be true for any layperson as well, of course, and clergymen did not become or remain clergymen on the basis of any externally imposed test of piety. Yet their lay peers would have measured them carefully against contemporary norms of piety, whether for breach or observance, making it a factor in their reputations. A clergyman's own sense of how he met or fell short of those expectations, or how they differed from his own, would shape his clerical identity.

One really interesting question gender historians can address, then, is what aspects of a masculine identity piety itself either enhanced or diminished. The answer will depend utterly on historical circumstances; no presentistic assumptions or bland generalizations will do. (Even today, in countries where women dominate everyday church life and have the most contact with clergy, the most important corporate expressions of piety, such as patronal processions, may be customarily reserved for men.) That opens up the possibility of considering to what extent clerics themselves used a gendered measure to criticize piety among their own male peer groups. I am wondering here how specific to the mid-twentieth century the experience of the young Alan Bennett was. In the 1940s Bennett, his theatrical career not yet on the horizon, was extremely pious, a 'regular communicant who knew the service off by heart' and disapproved of those who did not. But the young man perceived little approval of his piety from the Anglican clergy of Leeds, who he feared instead found it 'faintly embarrassing': 'Shy, bespectacled and innocent of the world I knew I was a disappointment to the clergy. What they wanted were brands to pluck from the burning, and that was not me by a long chalk; I'd never even been near the fire.'[26] Clearly, Bennett perceived that he reflected back to the clergy exactly the image of themselves they were trying to escape, the dainty Victorian caricature, and that they preferred a more conventionally masculine kind of Christian who would serve to attract conventionally masculine males into the church – thus serving the ends of one kind of piety. How did the gendering of piety weigh on clerics of earlier eras, we might ask? Once we attempt to answer that question we will have a much more complete sense of what 'clerical masculinity' actually meant.

VI   Some closing questions

In conclusion, I would like to point toward a couple of specific questions on which the concerns of masculinity history, premodern history and religious history come together. I have already mentioned the importance of the Protestant Reformation in defining the terms by which we conceive clerical masculinity. Now I would suggest that an attention to clerical masculinity itself – as discursive trope and as lived identity – would in turn illuminate our understanding of the Reformation itself. The terms of Reformation studies, particularly of the English Reformation, were set well before masculinity became a matter of concern even to gender historians. Gender in the Reformation has been studied with respect to women. The question of clerical masculinity would add to this body of work a more subtle understanding of anticlericalism, for example, whether with respect to sexuality and marriage, or otherwise. A fuller picture of the meaning of gender in the social compass of the Reformation would in turn help us to re-evaluate the status of the Reformation in conventional periodization, a contribution that would apply well outside the subdiscipline of gender history.

Clerical masculinity also allows historians to help refine the theory of masculinity studies, particularly the contrast between hegemonic, patriarchal, antipatriarchal, subordinate and alternative forms or patterns of masculinity that has so influenced the discipline. Should we classify clerical masculinity as 'hegemonic' because clerics were part of a hegemonic and normative institution, the Church? Should we consider it 'patriarchal', since in some ways the Church was the very model of patriarchal authority – even where priests were celibate and would never have legitimate children? Was it 'antipatriarchal' or 'alternative' because clerics may not have conformed to all behavioral codes that defined masculinity for some laymen? Or are even these labels themselves somehow unsuitable? (Confessors, and priests with cure of souls, for example, exercised a unique kind of power over their parishioners that had no analogue in lay society, but which was also out of the reach of clerics in lesser orders.) These are matters that will require some hard research and equally hard thinking, but both gender studies and historical studies will be the better for it.

Notes

1. John Tosh, 'What Should Historians Do with Masculinity? Reflections on Nineteenth-Century Britain', *History Workshop Journal* 38 (1994): 179–202.

2. The literature on marriage in the Reformation is vast, and here I note only a few studies that specifically address the clergy. Steven F. Ozment touched on the subject in his classic *When Fathers Ruled: Family Life in Reformation Europe* (Chicago, 1983), 1–24, though his discussion concentrates more on the implications for women. See also Scott H. Hendrix, 'Masculinity and Patriarchy in Reformation Germany', *Journal of the History of Ideas* 56 (1995): 177–93. While mostly concerned with the English situation, Helen L. Parish, *Clerical Marriage and the English Reformation: Precedent, Policy and Practice* (Aldershot and Burlington, 2000) gives a very useful overview of the principles involved in the Reformation debate, particularly the use English polemicists made of Continental ideas; in particular, see chapter 2, 'Celibate Priesthood or Married Ministry? The Testimony of the Bible' (pp. 39–65) to understand the arguments about scriptural authority.

3. This is well known, but for examples covering a variety of historical moments and contexts, see Rachel J. Weil, 'The Politics of Legitimacy: Women and the Warming-Pan Scandal', in *The Revolution of 1688–89: Changing Perspectives*, ed. Lois G. Schwoerer (Cambridge and New York, 1992), esp. 75–6; Peter Wagner, 'Anticatholic Erotica in Early Modern England', in *Erotica and the Enlightenment*, ed. Wagner (Frankfurt, 1991); Denis Paz, *Popular Anti-Catholicism in Mid-Victorian England* (Stanford, 1992), esp. 275–76; Susan David Bernstein, *Confessional Subjects: Revelations of Gender and Power in Victorian England* (Chapel Hill, N.C., 1997), 47–9.

4. While the simplest Internet search will turn up many popular expressions of this belief, for a more scholarly analysis see Mary Gail Frawley-O'Dea and Virginia Goldner, eds., *Predatory Priests, Silenced Victims: The Sexual Abuse Crisis and the Catholic Church* (London and New York, 2007), in particular Goldner's introductory essay 'The Catholic Sexual Abuse Crisis: Gender, Sex, Power, and Discourse', 1–20.

5. See, for example, Stanley Brandes, *Metaphors of Masculinity: Sex and Status in Andalusian Folklore* (Philadelphia, 1980), 177–204.

6. This 'Special Feature on Masculinities', in *Journal of British Studies* 44/2 (2005), 274–362, consists of six articles by Karen Harvey, Alexandra Shepard, Michèle Cohen, John Tosh and Michael Roper.

7. An early attempt to explore this question, now needing to be revisited, is Lyndal Roper, 'Was There a Crisis in Gender Relations in Sixteenth-Century Germany?', in *Oedipus and the Devil: Witchcraft, Sexuality and Religion in Early Modern Europe*, Lyndal Roper (London, 1994), 37–52.

8. Brian Thompson, *Keeping Mum: A Wartime Childhood* (London, 2006), 6.

9. Francis L. Bushman, *The Refinement of America: Persons, Houses, Cities* (New York, 1992), 313–52.

10. Jennifer D. Thibodeaux examines how this played out in social reality in 'Man of the Church, or Man of the Village? Gender and the Parish Clergy in Medieval Normandy', *Gender and History* 18/2 (2006): 380–99.

11. Anonymous, *My Curates, by a Rector* (1890; available from Project Canterbury, http://www.anglicanhistory.org/levity/curates/chapter8.html).

12. Early work on medieval masculinity made assumptions like this: JoAnn McNamara, 'The *Herrenfrage*: The Restructuring of the Gender System, 1050–1150', in *Medieval Masculinities: Regarding Men in the Middle Ages*, ed. Clare A. Lees (Minneapolis, 1994), 3–30; R.N. Swanson, 'Angels Incarnate: Clergy

and Masculinity from Gregorian Reform to Reformation', in *Masculinity in Medieval Europe*, ed. D.M. Hadley (London, 1999), 160–77.

13. For example, David Gilmore, *Manhood in the Making: Cultural Concepts of Masculinity* (New Haven, Conn., 1990).

14. I have consistently argued this methodological point in my own work; see particularly Derek Neal, 'Suits Make the Man: Masculinity in Two English Law Courts, c. 1500', *Canadian Journal of History* 37 (2002): 1–22; Neal, 'Masculine Identity in Late Medieval English Society and Culture', in *Writing Medieval History*, ed. Nancy F. Partner (London, 2005), 171–88; and Neal, *The Masculine Self in Late Medieval England* (Chicago, 2008).

15. I will not attempt here to provide references to the vast earlier literature on lay–clerical relations that predates the development of gender history. Notable recent studies (most do not explicitly address masculinity, but consider relevant issues) include the following: Joan Greatrex, 'Monk Students from Norwich Cathedral Priory', *English Historical Review* 106 (1991): 555–83; Greatrex, 'Rabbits and Eels at High Table: Monks of Ely at the University of Cambridge, c. 1337–1539', in *Monasteries and Society in Medieval Britain: Proceedings of the 1994 Harlaxton Symposium*, ed. Benjamin Thompson (Stamford, U.K., 1999), 312–28; Virginia Davis, 'Preparation for Service in the Late Medieval English Church', in *Concepts and Patterns of Service in the Later Middle Ages*, ed. A. Curry and E. Matthew (Woodbridge, U.K. and Rochester, N.Y., 2000), 38–51. See also Katherine L. French, *The People of the Parish: Community Life in a Late Medieval English Diocese* (Philadelphia, 2001). P.H. Cullum raises the gender implications in 'Clergy, Masculinity and Transgression in Late Medieval England', in *Masculinity in Medieval Europe*, ed. Hadley, 178–96, and adds the crucial element of life stage in 'Boy/Man into Clerk/Priest: The Making of the Late Medieval Clergy', in *Rites of Passage: Cultures of Transition in the Fourteenth Century*, ed. Nicola McDonald and W. Mark Ormrod (York, 2004), 51–66. Most recently, note Thibodeaux, 'Man of the Church?'.

16. Again, this theme is treated in much earlier social history and especially Reformation history. Among more recent studies see R.L. Storey, 'Malicious Indictments of Clergy in the Fifteenth Century', in *Medieval Ecclesiastical Studies in Honour of Dorothy M. Owen*, ed. M.J. Franklin and Christopher Harper-Bill (Woodbridge, Eng., 1995), 221–40; Swanson, 'Angels Incarnate'; Cullum, 'Clergy, Masculinity and Transgression'; Thibodeaux, 'Man of the Church'; Neal, 'Husbands and Priests: Masculinity, Sexuality, and Defamation in Late Medieval England', in *The Hands of the Tongue: Essays on Deviant Speech*, ed. Edwin D. Craun (Kalamazoo, Mich., 2007), 185–208; Neal, *The Masculine Self*.

17. Dyan Elliott, 'Pollution, Illusion, and Masculine Disarray', in *Constructing Medieval Sexuality*, ed. Karma Lochrie, James A. Schultz and Peggy McCracken (Minneapolis, 1997), 1–23; Conrad Leyser, 'Masculinity in Flux: Nocturnal Emission and the Limits of Celibacy in the Early Middle Ages', in *Masculinity in Medieval Europe*, ed. Hadley, 103–20; Jacqueline Murray, 'Mystical Castration: Some Reflections on Peter Abelard, Hugh of Lincoln and Sexual Control', in *Conflicted Identities and Multiple Masculinities: Men in the Medieval West*, ed. Jacqueline Murray (NY, 1999), 73–92; Murray, ' "The Law of Sin That Is in My Members": The Problem of Male Embodiment', in *Gender and*

Holiness: Men, Women, and Saints in Late Medieval Europe, ed. Sarah Salih and Samantha Riches (New York, 2002), 9–22.

18. Nancy F. Partner, 'The Family Romance of Guibert of Nogent: His Story/Her Story', in *Medieval Mothering*, ed. John Carmi Parsons and Bonnie Wheeler (New York, 1996), 359–79; Brian Patrick McGuire, 'Sexual Awareness and Identity in Aelred of Rievaulx (1110–67)', *The American Benedictine Review*, 45/2 (1994), 184–226; McGuire, 'Jean Gerson and Traumas of Male Affectivity and Sexuality', in *Conflicted Identities*, ed. Murray, 45–72.

19. Recognized also in Karen Harvey and Alexandra Shepard, 'What Have Historians Done with Masculinity? Reflections on Five Centuries of British History, Circa 1500–1950', *Journal of British Studies* 44/2 (2005): 274–80, esp. 275–6.

20. A distinction originally elaborated by R.W. Connell in *Masculinities* (Berkeley and Los Angeles, 1995), chapter 3, and most recently revisited in R.W. Connell and James W. Messerschmidt, 'Hegemonic Masculinity: Rethinking the Concept', *Gender and Society* 19/6 (2005): 829–59.

21. Treated both in Neal, 'Husbands and Priests', and at greater length in Neal, *The Masculine Self*.

22. In which case, the work of Ruth Mazo Karras on university students is very pertinent: see her 'Separating the Men from the Goats: Masculinity, Civilization, and Identity Formation in the Medieval University', in *Conflicted Identities*, ed. Murray; and her *From Boys to Men: Formations of Masculinity in Late Medieval Europe* (Philadelphia, 2003).

23. See F. Donald Logan, *Runaway Religious in Medieval England c. 1240–1540* (Cambridge, 1996), for a vivid account of these pressures.

24. See particularly Jacqueline Murray, 'Masculinizing Religious Life: Sexual Prowess, the Battle for Chastity and Monastic Identity', in *Holiness and Masculinity in the Middle Ages*, ed. P.H. Cullum and Katherine Lewis (Toronto and Buffalo, 2005), 24–42. See also John H Arnold, 'The Labour of Continence: Masculinity and Clerical Virginity', in *Medieval Virginities*, ed. Anke Bernau, Ruth Evans and Sarah Salih (Cardiff, 2003), 102–18.

25. The recent collection, *Holiness and Masculinity*, ed. Cullum and Lewis, contains a number of essays addressing this subject, especially Christopher C. Craun, 'Matronly Monks: Theodoret of Cyrrhus' Sexual Imagery in the *Historia Religiosa*' (43–57) and Carolyn Diskant Muir, 'Bride or Bridegroom? Masculine Identity in Mystical Marriages' (58–78).

26. Alan Bennett, *Talking Heads* (first publ. 1988; London, 1997), 17–8.

Part I

Monastic Masculinity: Bridging Secular and Clerical Identities

2
The Common Bond of Aristocratic Masculinity: Monks, Secular Men and St. Gerald of Aurillac

Andrew Romig

Miracle stories had been swirling throughout the densely forested hills south of the Loire for a decade when, sometime during the early 920s, Abbot Aymo of St. Martial called upon his esteemed friend, Odo of Cluny, to discern whether they might actually be true. Like most tales of their kind, they spoke of spontaneous healings – healings of blindness in particular, but also of deafness, paralysis and demonic possession. They told of a man who had lived a life of exemplary love and devotion toward his fellow neighbors and the Christian God. Yet familiar as the stories might have seemed in form, there was something decidedly troubling about them in content. Aymo needed a man of Odo's learning and gravitas to investigate because the alleged holy man at the stories' center had lived his life not in the formal service of God, but instead as a wealthy and powerful layperson.[1] He had been Count Gerald of Aurillac, warlord and secular judge of the fragmented political world of the Carolingian late ninth and early tenth century, who had died of old age in the year 909.[2]

Odo initially presumed that the tales were false.[3] Odo believed fully in the monastic commonplace that defined the secular world as a source of virtually inevitable corruption for the soul.[4] And besides, there had never been an officially recognized lay miracle worker who had not also borne in life the special aura of a king or died as a martyr.[5] Thus when, after a long and careful investigation, Odo declared that Gerald of Aurillac was in fact a saint, and that he had indeed performed both his saintly miracles and his comital duties at the same time, it could not have been a conclusion that he reached lightly. We must view Odo of Cluny's *De vita Sancti Geraldi Auriliacensis comitis Libri Quattuor* – the *vita* that he wrote to commemorate the man – as a startling pronouncement from an important medieval mind about the possibilities of sanctity and the access of secular men to divine power within the Christian community.

Precisely what kind of pronouncement Odo was making, however, has been the subject of considerable scholarly debate. Historians have puzzled over the fact that the *Vita Geraldi* seems to convey a decidedly mixed message about the ideal of Christian living that it espouses. Odo wrote expressly that St. Gerald was to serve as a model for 'the mighty' (*potentes*) of his world.[6] Yet Gerald shuns nearly all of the traditional pursuits of aristocratic Frankish secular men in favor of monastic virtues. Gerald does not live in the opulence normally associated with his station, but rather shares his great wealth with the less fortunate. He does not continue his family line through marriage and procreation, but rather lives a life of strict chastity. He does not fight as a warrior, but rather shuns killing, never even sheds blood and famously orders his men to do battle with the flats of their swords and blunt ends of their spears.[7]

These incongruities have led most scholars to see in Gerald's character the evidence of irresolvable secular and clerical identity conflict within Carolingian culture. For Jacqueline Murray, the *Vita Geraldi* demonstrates 'the extent to which the worldly life of action and the imperatives of the religious life were perceived to be at odds.'[8] Stuart Airlie has written that, while the *Vita Geraldi* ostensibly confronted 'head on' the issue of how to reconcile Frankish secular life with Christian duty, 'Odo's intensely monastic preoccupations made his Gerald an inappropriate model.'[9] And Janet Nelson has suggested that the *Vita Geraldi* records nothing less than an outright crisis of masculine identity in the latter half of the ninth century, in which aristocratic Frankish boys were forced to negotiate impossibly 'competing, even conflicting, models of manliness', and could face mental anguish and even physical sickness as a result.[10]

In this essay, I reopen the discussion of this well-known text, not to argue against these assessments so much as to qualify how we conceive the relationship between secular men and men 'in religion' during the Carolingian period. Odo's text has been of such great interest to scholars because of its depictions of the secular world, yet the focus on St. Gerald as a model for secular men has obscured the fact that he served Odo as a model not just for laymen, but for monastic men as well.[11] Embedded within the *Life of St. Gerald* is a local message, a not particularly subtle barb that Odo aimed directly at the monks of Gerald's region, whose laxity and questionable masculinity, in Odo's judgment, had led to Gerald's appearance in the first place. It was a critique that made

Gerald a paragon of the ideal masculinity that both groups were to share in common.

It should be said that in my analysis I choose to view the *Vita Geraldi* in a more literary light than scholars previously have. There is no doubt a great deal of historical truth in Odo's account – enough to satisfy the readers who actually knew the man in life. But I am frankly far less concerned with Gerald the historical figure than I am with Odo's comprehension and representation of him. Whatever else the text may be, the *Vita Geraldi* remains at its heart a narrative of how Odo saw, understood, explained and then naturalized a form of Christian sainthood that did not conform to any of the more fixed models that his culture normally recognized. My hope is to enhance our view of how Odo framed the masculinity of this saintly secular man and thus to explore the kinds of work that the *Life of St. Gerald* performed in defining male gender for all contemporary aristocratic men – monastic as well as lay.

I   The common bond of Christian men in Carolingian thought

As scholarship of the *Vita Geraldi* has suggested, Gerald no doubt caused Odo to recall conceptions of ideal secular Christianity that had developed during the century before his own, when ecclesiasts preached the universality of the Christian soul and a common bond among all Christian Frankish men. Ninth century thinkers founded their models of lay Christianity upon a monastic ideal, so much so that Smaragdus of St. Mihiel could adapt with very little change his treatise on the monastic life, the *Diadema monachorum*, into a 'mirror' for Emperor Louis the Pious.[12] Heeding the best scientific knowledge of their day, they knew that ascetic self-denial – humility, abstinence from wealth and sexual restraint – was the best means for men to maintain the heat, dryness and continence that defined the ideal male body.[13]

The apparent incompatibility between these monastic values and Frankish aristocratic culture has led scholars to assume a corresponding lack of connection between Carolingian spiritual advice and the realities of secular life in that era.[14] But such an assessment fails to recognize in full the extent to which this ninth-century advice sought expressly to transcend difference and to profess the essential sameness of all Christian souls. Spiritual advisors affirmed, over and over again, the fundamental equality of secular and 'professional' religious men in the eyes of God, and the common call to public service and good works that they both shared together.

In a public epistle to a certain Count Wido of Brittany, Alcuin of York argued for the equivalency of the secular and clerical soul.[15] 'Let not the nature of your lay habit or of your secular association frighten you', he urged the count in a remarkable passage,

> as though in this garb you will not be able to enter the gateway of celestial life. For just as the blessedness of the kingdom of God is preached to all equally, so is the entrance to the kingdom of God open to all sexes, ages, and persons equally according to the worthiness of their merits. For there, there is not the distinction which there is in the secular world between layman and cleric, rich man and poor man, young man and old, slave and master. Rather, each will be crowned in perpetual glory according to the merit of their good work.[16]

Words such as these from Alcuin were certainly a means of encouraging lay Christians toward the service of God and, no doubt, the monetary support of the Church. But they were also a means of creating a sense of common Christianity which, as Thomas F.X. Noble has recently suggested, evolved quickly into an ethos of nobility for the entire Carolingian aristocratic caste.[17] They inspired laypeople to understand that they, too, had not only the capacity but also the duty to comprehend and to perform the same good deeds demanded of the clergy.

The effect was to perforate the perceived boundaries that separated lay and 'professional' Christian identities. Perhaps the most expansive discussion of this theme of commonality during the ninth century came in a Book of Exhortation, written by Bishop Paulinus of Aquileia to Count Henry of Friuli, but disseminated widely after the 820s.[18] The bishop stressed that all persons, regardless of rank or way of life, were to live by the same Christian laws decreed by God. 'Be, I seek', Paulinus wrote, 'although a layman, prompted to all work of God, kind to the poor and sick, consoler to the dying, compassionate to the miseries of all people....'[19] There was an equal moral burden placed on all Christians, he said, regardless of the particular walk of life from which they hailed. Paulinus reminded Henry that God promised the Kingdom of Heaven not only to 'us', referring to himself as a member of the clergy but also to all laypersons who serve God's precepts 'with their whole heart'.[20] Christ poured out his blood, Paulinus warned, 'not only for us clerics, but also for the whole human race, who are predestined for eternal life'.[21] There was a 'great confusion' among laypeople who believed that 'a clergyman...should do the things that a clergyman does' and that laypersons by extension were exempt from service; such people did not understand that in order to share in the earthly goods that God provided and the happiness of the Kingdom of Heaven, all had 'to carry the yoke of Christ with equal labor' (*aequali labore jugum Christi ferre*).[22]

Paulinus and Alcuin echoed the philosophies of Christian fathers such as Isidore of Seville and Gregory the Great in locating difference as a function of the interiority and exteriority of the human being.[23] Paulinus wrote that the inner self was far more important than the outer in the eyes of God, because Henry's 'interior man' (*interior homo*) bore the image of God ('the Builder') himself. Inside the body, the intellect, will and memory all imitated the Holy Trinity.[24] God added 'the grace of spiritual knowledge' to Henry's head so that it might lead him toward eternal life. Thus Henry had to use this knowledge to control his will and to maintain his sights on last things and perpetual beatitude.[25] Paulinus again returned to the theme of the 'interior man' in an extended discussion of good work, describing how Henry was to override his own (external) pleasures and instead cultivate love of God in his mind and 'heart'.[26] 'Let us have the love of God and of neighbor inside us', Paulinus urged, for from love came tranquility: 'He who is filled with love walks with tranquil soul and most serene face.'[27] For laypeople, just as for clergymen, Paulinus wrote, good work flowed from an alignment of the interior will away from the secular world and toward God and salvation: 'If we desire to possess anything in this secular world, let us possess with unencumbered mind God, who possesses all things, and let us hold in him whatever we happily and in a holy manner desire.'[28]

Around the year 820, Bishop Jonas of Orléans wrote a treatise on the ideal lay life, the *De institutione laicali*, in response to the petition of another count, Matfrid of Orléans, who wanted to learn more about how best to live as a Christian in the world.[29] Jonas was familiar with the works of both Paulinus and Alcuin on the subject and he tailored his text to resemble their works.[30] Jonas' treatise did not blur all boundaries of identity: it focused on marriage as the primary feature that separated the layperson's way of life from that of the clergy.[31] Still, far more important for the bishop of Orléans were the obligations to God that laypeople and clergy shared in common. In an even more strident tone than Paulinus and Alcuin, Jonas made sure to explain that both laity and clergy alike were held to the same commandments of God. Jonas explained to Matfrid that clergy and laity were equal in their combined obligation to follow love (*caritas*) as the highest principle of the Christian life.[32] 'The law of Christ is attributed by the Lord not specially to clerics', he wrote, 'but it is to be observed generally by all the faithful'.[33] Echoing Paulinus' admonition to Count Henry, Jonas warned Matfrid that he should not attempt to emulate those laypeople who believed, falsely, that they could be saved by their faith alone, and who claimed that the precepts of God therefore pertained only to clerics and not to them.[34] This was folly, he said, for the laypeople he knew managed all the time to believe that the law handed down by God was given to them. They managed to strive, just like their clerical brethren, to live by that law to the best of their ability. All sinners – lay and cleric alike – had to perform good works of love in order to achieve the Kingdom of Heaven.[35]

For Jonas, therefore, the common bonds of original sin and salvation, and the subsequent obligation to God that all Christians shared equally, were of far greater import than any lines of worldly identity that might separate them. Jonas did recognize the call to ascetic chastity and bodily purity as an important distinction between secular Christians and cloistered monks. Nevertheless, he did not privilege the ascetic life over the worldly orders of the Church. He simply acknowledged that life in the cloister and in hermitage was subject to a set of commandments that did not otherwise apply to those living in the world. 'Although', he wrote, 'in the Gospel there are certain special precepts which are only appropriate for despisers of the world and emulators of the apostles, the rest are decreed indiscriminately without pretext to all the faithful, each of course according to the order by which one vows to serve God.'[36] The sole and primary commandment for all Christians was still to love God and one's fellows indiscriminately. Guided by that love, all members of the Church, regardless of their chosen way of life, were required to follow God's precepts and to carry out good works. All had not only the capacity but also the responsibility to become elite disciples to the greatest extent that their free will allowed.

These treatises demonstrate the existence of an ideal of Christianity during the ninth century through which spiritual advisors emphasized the equality of both laypeople and clergy in the eyes of God. This ideal recognized the different duties and uses of the body that separated secular from priestly and monastic ways of life. But the perpetual focus remained the common bonds that made all Christians part of one *ecclesia*. All walks of Christian life required sober recognition of the corrosive forces of the world and a stern refusal to allow those forces to distract the soul from its pursuit of salvation. Anyone who dared to live within the harsh confines of the *saeculum* had to be aware of the constant threat that it posed. But this threat was dire regardless of a person's chosen path. Whether one's outer self wore the garb of the Church or the garb of the secular world mattered not at all in the eyes of God.

II   Miracles, men and manliness in the Auvergne

We must read Odo's *Vita Sancti Geraldi* as an application of this ideal of essential equality, shared capacity and common obligation between lay and 'professional' Christians. Odo clearly believed that Gerald's secularity was potentially dangerous. In the *vita*'s preface, he alluded derisively to a significant group of laypeople who, instead of marveling at how Gerald was able to achieve sainthood as a worldly man, used Gerald's lay status as justification for their sinful worldly excesses: 'They strive indeed to excuse their luxurious lives by his example', Odo accused.[37] Yet Odo also believed that the world could be precisely the place where monks might look for inspiration and for models of heroism in the service of God. It was the true battlefield on which the virtues and vices fought for the spoils of human souls, and life there demanded a degree of mettle and manliness that was beneficial for salvation and that the cloister could in fact corrupt.[38] Odo therefore had to perform the tricky balancing act of acknowledging Gerald's saintly secularity while also rendering it unthreatening to ascetics such as himself, whose way of life was more austere and whose traditionally exclusive access to the heights of elite Christian discipleship had to be jealously guarded. The result was a presentation of Gerald as an amalgam of secular and monastic forms.

Outwardly, of course, the two sides of Gerald's character appear to coexist in harmony in the *vita*. He is born the son of a prominent Frankish aristocratic family and learns all of the values, duties and pursuits of his station; his parents provide him with a proper secular education, which included training in horsemanship, archery, hunting and falconry.[39] As a count, he leads his armies against rival warlords and adjudicates criminal complaints.[40] Yet, following traditionally monastic virtues, he is always a humble champion of those beneath him.[41] He regularly takes his meals with the less fortunate; he gives alms, food and clothing to whoever is in need; and he never turns away the destitute beggars who know that they may come to his door for a meal whenever they are in need.[42] Odo wrote that Gerald balanced admirably his worldly duties as a count with his religious obligations to God. He stepped 'neither to one side nor the other from the *medioximum callem* (the 'middlemost path') of discretion, so that he neither failed in the duties of his worldly affairs, nor diverted himself from the practice of religion by his earthly occupations'.[43] 'Rightly he was loved by all', the *vita* says, 'for he himself loved everyone'.[44]

The rest of Odo's narrative, however, belies this outer concordance, for Gerald must continually work against secular masculine norms.[45] In conformity with monastic chastity, Gerald exhibits a decidedly cool attitude toward women: he is 'saved' from taking a wife by God himself, who renders the girl in whom Gerald takes an interest suddenly deformed and hideously ugly to the young count's eyes; God then strikes him blind for more than a year so that he will not be tempted again by female beauty, and he lives the rest of his life as a celibate.[46] But Gerald's negotiation of his sexuality actually receives far less attention in the *vita* than his negotiation of his duties as a count. He must constantly work alone, often in opposition to the very power structures over which he himself presides.[47] The most famous parts of the text demonstrate furthermore the heroic, indeed miraculous, cartwheels that Gerald must perform in order not to kill. Odo wrote that Gerald strove to repress violence by promising peace and offering easy reconciliation to his enemies.[48] But when this did not work, 'led by love of the poor, who were not able to protect themselves', he would order his men to fight with the butts of their swords and reversed spears – a command that would have been ridiculous, Odo continues, had not God made Gerald and his men invincible before the enemy.[49] This 'new kind of doing battle mixed with lovingness (*pietas*)' was what allowed Odo to silence the skeptics.[50] 'Let no one be worried', he argued, addressing his monastic audience directly, 'because a just man sometimes made use of fighting, which seems incompatible with religion. No one who has judged impartially will be able to show that the glory of Gerald is clouded by this.'[51]

In these episodes, the inner conflict that Gerald experiences appears at first to be rather one-sided. Odo depicts a Gerald who challenges and rebels against the norms of secular society, nurturing and championing the monastic side of his identity. But as in the ninth-century texts discussed above, Odo is reluctant to draw an impermeable line between secular and monastic men. The stories about Gerald increasingly reveal how the count looks upon *both* his lay and monastic fellows as strangers.

Gerald would have fled the secular world entirely had not the cloister proved to be just as flawed. Book Two demonstrates clearly how Gerald begins to live quite literally as a monk in lay clothing. Odo explains in detail the way in which Gerald takes the tonsure in secret, hiding it from all but his chamberlains through a clever placement of his crown. He wears the linen vestments of the monastic life next to his skin, under the richer leather of his secular uniform. He rides horses, but he does not adorn them with the finery associated with his class. And though he continues to wear his sword and scabbard to keep up his outward secular appearances, he never touches them again.[52] Importantly, however, Gerald is never quite able to leave his post as count and join a monastic community, for he is unable to find any men manly enough to join him in the common life.

Odo wrote that Gerald donated his possessions to God and consecrated properties in Aurillac for the establishment of a monastery, 'For he very much desired to establish a monastic foundation in that place, where the monks might lead the common life with an abbot of their order.'[53] The story continues to relate, however, that even though Gerald begins construction with great zeal, unstable foundations frustrate his plans. The instability is structural: after he spends a great deal of money, the building collapses because of shoddy workmanship.[54] And it is metaphoric: Gerald rebuilds the new monastery from the ground up, but then searches in vain for 'monks of good character' (*monachos qui bene morati essent*) to live there. When he finds none, he decides to create a group of good men on his own, sending a collection of young boys of the noble class away to learn the monastic rule. These monks, too, however, when they return to live in Gerald's monastery, quickly languish in what Odo tellingly labels 'girlish softness' (*puellari mollitia*).[55]

Feminine softness is briefly mentioned earlier in the *vita* when Gerald's soldiers complain, in the context of the count's remarkable kindness and zealous peacemaking, that his actions make him appear 'soft' (*mollis*) and 'timid' (*timidus*).[56] But in that case, Odo himself answers the secular challenge to Gerald's manliness through the vindication of Gerald's example: in the *vita*, God demonstrates repeatedly that Gerald's leniency is just and indicative of strength rather than weakness. The only explicitly gendered criticism that Odo himself makes in the *Vita Geraldi* is against monasticism, and Odo did not stop there. The *vita* continues to describe how Gerald never forgets his desire to congregate a community of monks and often speaks of it with his friends. 'O', laments the count in direct speech,

'if it might be granted me by some means to obtain religious monks. How I would give them all I possess, and then go through life begging. I would make no delay in taking the necessary steps.' Sometimes his friends would say, 'Are there not many monks to be found in these regions from whom you can choose a community at will?' But speaking with great vehemence he would reply, 'If monks are perfect, they are like the blessed angels, but if they return to the desire of the world they are rightly compared to the apostate angels, who by their apostasy did not keep to their home. I tell you that a good layman is far better than a monk who does not keep his vows.' When they rejoined, 'Why then have you been accustomed to show such favors not only to neighboring monks, but to those from a distance?' making little of his deeds, with his usual humility he would reply, 'What I do is nothing; but if, as you say, I do anything, I am certain that He is true who promised to reward a cup of cold water given in His name. Let them understand what they are in the eyes of God. Certainly it is true that he who receives a just man in the name of a just man shall receive the reward of a just man.' These and similar words of his make it clear that he despised the pleasures of this present life, that he burned with the desire of heaven, that he wished to leave all his possessions, if there had been anyone to whom he could reasonably hand them over .... It was against his will, therefore, that he was kept in the world. And although companions were lacking with whom he might renounce the world, he occupied himself entirely in a wonderful way with the work of God.[57]

This conversation between Gerald and his friends reveals a crucial, and largely overlooked, message that Odo sought to convey through the *Vita Geraldi*, a thinly veiled admonition from Odo that would have held far more meaning for a monastic reader than a secular one: 'A good layman is far better than a monk who does not keep his vows', Odo says in Gerald's voice. It is a statement that holds St. Gerald up as an *exemplum* not just for secular men, but for monastic men as well, presenting the kind of masculine virtue that all aristocratic men could achieve, and that the professional religious men of Gerald's region lacked. It also happens to offer a compelling explanation for why the anomaly of Gerald had occurred in the first place. Gerald was a secular saint, and not a monastic one, because monasticism in the Auvergne had grown cold. The count would otherwise have become a monk, suggests Odo, had there been anyone man enough to join him.

III   Masculinity in the age of Odo of Cluny

With this explicitly gendered critique of the professional religious men of the Auvergne, the passage highlights the localism of the history that the *Vita Geraldi* contains within. But more broadly, it allows us to see a conception of masculinity at work during the tenth century that could both encompass and transcend the boundaries of secular and monastic identity. It was an ideal represented by the virtuous example of St. Gerald: not just of monastic Christian celibacy and non-violence, not just of secular Christian protection of the poor and founding of churches, but of universally masculine vigor, striving and tireless vigilance against vice.

The history of medieval masculinity commonly focuses on the oppositions between secular and monastic ideals and the resulting competitions between laymen and clerics for hegemony over cultural definition of gender. There can be no doubt that the different values of the two ways of life produced rivalry, but this rivalry did not only lead to contest and anxiety over difference. Devout laypeople tried as well as they could to live in accordance with New Testament precepts; monks fashioned themselves as the virile warriors of Christ.[58] But such negotiations were not just imperfect compromises that grew out of social and personal pressures. They stemmed from the fact that warriors and monks could consider themselves, at essence, to be more similar than different, bound to follow a common masculine ethos.

Rather than reading the *Vita Geraldi*, therefore, as evidence of the difficulties of translating monastic ideals into the secular world, we should see it as continuing to remind both secular men and men given to the religious life of their common bonds not just as Christians but as men of the Frankish nobility. It demonstrates the capacity of Odo and his audience to hold various forms of masculinity in tension without the need for either the prescriptive reconciliation of that tension or the conception of new gender categories to accommodate differences. It is important, in this light, to entertain the distinct possibility that Odo may never have meant for St. Gerald to represent the new ideal of lay Christianity that modern scholars have ascribed to him. The simple fact that Gerald's hybridity could only be achieved truly through miracle rendered him safely irreproducible, and was precisely what allowed Odo to laud a secular saint without ever compromising the monastic values that he himself held dear as part of his own self-identity. Nevertheless, the reading of the text that I have presented here forces us to shift our comprehension of the relationship between monastic and secular men in this era. Rather than seeing them as competing against one another for control over a single overarching masculine embodiment, we must describe them as sharing two sides of the same gender coin, different but inextricably linked and therefore mutually capable of claiming maleness without undermining the masculinity of the other.[59]

Finally, a text such as the *Vita Geraldi* might even force us to reorient how we understand masculinity itself during this period. It is entirely possible that we focus too heavily on the bodily markers of masculinity upon which we dwell in contemporary culture to the exclusion of markers of masculinity that early medieval men historically held most dear. Caroline Walker Bynum remarked 20 years ago that the focus in our modern historical work on sex and money as the driving forces of society and culture may be far more a reflection of our own world than of the Middle Ages.[60] Could it be, for example, that in the tenth century, sexuality was seen as a far less important marker of masculinity than we presume? Could it be that sexuality was a marker of worldly life, but not necessarily a particular marker of manliness? Could it be that the far more important performances of male gender in this culture were performances of service, hard work and striving for perfection? Both the ninth-century treatises on the ideal lay life and Odo's tenth-century depiction of a lay saint would certainly suggest an affirmative answer to all of these questions. Regardless of exterior habits, monks and warriors were all equally to be men of the Frankish aristocratic caste.

Notes

1. Odo, who had once been a student of the famous Master Remigius (d. 908) at Auxerre, was counted among the most revered intellects of his day. See John Marenbon, *Early Medieval Philosophy (480–1150): An Introduction* (London and New York, 1988), 71–89, esp. 78–9. For Odo's learning, life and times, see most recently Christopher A. Jones, 'Monastic Identity and Sodomitic Danger in the *Occupatio* by Odo of Cluny', *Speculum: A Journal of Medieval Studies* 82, no. 1 (2007): esp. 5–10, with notes. See also Patrick Wormald, 'Aethelwold and His Continental Counterparts: Contact, Comparison, Contrast', in *Bishop Aethelwold: His Career and Influence*, ed. Barbara Yorke (Woodbridge, Eng., 1988), 19–22.
2. Gerald is a ubiquitous figure in scholarship of the tenth and eleventh centuries, though few studies treat him exclusively. For biographical information see Joseph-Claude Poulin, 'Geraldus Von Aurillac', in *Lexikon des Mittelalters* (Stuttgart, [1977]–1999).
3. Odo of Cluny, *De vita sancti Geraldi Auriliacensis comitis libri quatuor*, praefatio, in *Patrologia Latina* (PL) 133: 640. Translations, with minor tweaking from me, are from Gerard Sitwell, reprinted in Thomas F.X. Noble and Thomas Head, *Soldiers of Christ: Saints and Saints' Lives from Late Antiquity and the Early Middle Ages* (University Park, Pa., 1995), 295–362.

4. Cf. Odo of Cluny, *Collationum libri tres*, PL 133: 611–12. For comment about this sentiment in Odo of Cluny's other great work, the *Occupatio*, see Jan M. Ziolkowski, 'The *Occupatio* by Odo of Cluny: A Poetic Manifesto of Monasticism in the 10th Century', *Mittellateinisches Jahrbuch: Internationale Zeitschrift für Mediävistik* 24–25 for 1989–1990 (1991): 559–67.

5. For lay sanctity, see Joseph-Claude Poulin, *L'idéal De Sainteté Dans L'aquitaine Carolingienne, D'après Les Sources Hagiographiques, 750–950* (Québec, 1975), 82ff. Odo's *vita* is the earliest life of a non-royal lay saint attested in the manuscript tradition. Historians disagree, however, as to whether the anonymous *Vita Gangulfi*, another hagiographical account of a non-royal lay saint, might have been written during this period as well, perhaps even prior to Odo's life of Gerald. Its earliest textual attestation dates from the second half of the tenth century, and thus I follow the majority of scholarship, which places Odo's *vita* as the earliest; cf. I. Deug-Su, 'Note Sull'agiografia Del Secolo X E La Santità Laicale', *Studi medievali* 30, no. 1 (1989):143–62. For more on the *Vita Gangulfi*, see *Vita Gangulfi martyris Varennensis*, ed. W. Levison and Bruno Krusch, *Monumenta Germaniae Historica, Rerum Scriptores Merovingicarum* 7 (1919–20, Reprint 1979): 142–74. Levison states that the *Vita Gangulfi* appears in no fewer than 65 extant codices, in comparison to the *Vita Geraldi*, which is extant in 36.

6. Odo of Cluny, *Vita Geraldi*, praefatio; PL 133: 642: 'Quoniam vero hunc Dei hominem in exemplo potentibus datum credimus...'

7. See below for further discussion.

8. Jacqueline Murray, 'Masculinizing Religious Life: Sexual Prowess, the Battle for Chastity and Monastic Identity', in *Holiness and Masculinity in the Middle Ages*, ed. Patricia H. Cullum and Katherine J. Lewis (Cardiff, Wales, 2006), 24.

9. Stuart Airlie, 'The Anxiety of Sanctity: St Gerald of Aurillac and His Maker', *Journal of Ecclesiastical History* 43, no. 3 (1992): 375, 395.

10. Janet L. Nelson, 'Monks, Secular Men and Masculinity, c. 900', in *Masculinity in Medieval Europe*, ed. Dawn M. Hadley (London and New York, 1999), 123, 141–42.

11. We must remember that even if, as scholars have presumed, the *Vita Geraldi* enjoyed a significant lay audience during and after the tenth century, the majority of its readers would still have been monks. The most current assessment of the manuscript tradition supports this argument, as the earliest versions of the text appear to have monastic provenance. See Anne-Marie Bultot-Verleysen, 'Le Dossier de Saint Géraud d'Aurillac', *Francia* 22, no. 1 (1995): 176ff.

12. Smaragdus wrote his *Via regia* as an adaptation of the *Diadema monachorum*: Smaragdus of St. Mihiel, *Smaragdi abbatis diadema monachorum*, PL 102: 593–689; *Smaragdi abbatis via regia*, PL 102: 931–70. See, most recently, Matthew D. Ponesse, 'Smaragdus of St. Mihiel and the Carolingian Monastic Reform', *Revue Bénédictine* 116, no. 2 (2006): 367–92.

13. For an excellent recent discussion of ideals of the male body in the early medieval world, and their connection to classical conceptions of gender and the body, see Lynda L. Coon, 'Gender and the Body', in *Early Medieval Christianities, c. 600–1100*, ed. Thomas F.X. Noble and Julia M.H. Smith (New York, 2008).

14. 'Advice is given out on matters of importance to the secular aristocracy, but we do not see what difficulties the recipient might have in following it.' Airlie: 380, see also 376, 395. Cf. also Janet L. Nelson, 'On the Limits of the Carolingian Renaissance', in *Politics and Ritual in Early Medieval Europe* (London and Ronceverte, W. Va., 1986), 57. Rafaele Savigni has recently argued, in light of the apparent disconnection, that there were at least two different orientations with regard to laypersons in the ecclesiology of the ninth century: one ascetic-monastic, and therefore ultimately incompatible with secular life, the other more pragmatic, focusing on behaviors that were possible for laypeople to perform such as almsgiving and directed violence in defense of the non-warrior members of society. Raffaele Savigni, 'Les Laïcs dans l'ecclësiologie Carolingienne: Normes Statutaires et Idéal de "Conversion"', in *Guerriers Et Moines: Conversion Et Sainteté Aristocratiques Dans L'occident Médiéval, IXe–XIIe Siècle*, ed. Michel Lauwers (Antibes, 2002), 62. The sole exception to the general assumption of incompatibility comes from the late J.M. Wallace-Hadrill, who deemed the model of Christianity taught to laypeople during this period to have been 'all very practical'. J.M. Wallace-Hadrill, *The Frankish Church* (New York, 1983), 409. For our best guesses about ninth century secular Frankish culture, see especially Franz Irsigler, 'On the Aristocratic Character of Early Frankish Society', in *The Medieval Nobility: Studies on the Ruling Classes of France and Germany from the Sixth to the Twelfth Century*, ed. Timothy Reuter (New York, 1979); Pierre Riché, *La Vie Quotidienne Dans L'empire Carolingien* (Paris, 1973), 75–120. Cf. also Dennis Howard Green, *The Carolingian Lord; Semantic Studies on Four Old High German Words: Balder, Frô, Truhtin, Hêrro* (Cambridge, 1965).

15. Alcuin of York, *De virtutibus et vitiis ad Widonem comitem*, PL 101: 613–38. This text has never been translated; all translations herein are my own. Regarding the public distribution of this letter, Rosamond McKitterick notes that a manuscript of this text was apparently possessed by the monastery at Reichenau, appearing in its catalogue of texts dating 835–42; it also appears on the list compiled for Reginbert, which describes it as *Et duo libelli Alchuini ad Vitonem*. See the larger description of the text in Rosamond McKitterick, *The Frankish Church and the Carolingian Reforms, 789–895* (London, 1977), 168–70.

16. Alcuin, 36; PL 101: 638: 'Nec te laici habitus vel conversationis saecularis terreat qualitas…Igitur sicut omnibus aequaliter regni Dei praedicata est beatitudo, ita omni sexui, aetati, et personae aequaliter secundum meritorum dignitatem regni Dei patet introitus. Ubi non est distinctio, quis esset in saeculo laicus vel clericus, dives vel pauper, junior vel senior, servus vel dominus: sed unusquisque secundum meritum boni operis perpetua coronabitur gloria.'

17. 'Let us sum up this ideal, this noble ethos, which it might not be an exaggeration to call an ideology. It is remarkably simple and is solidly coherent with the central tenets of the Carolingian reform programme as those began to be articulated in the central years of Charlemagne's reign. Basically, it held that nobles must live an active public life, that they must accept the responsibilities that go with that life.' Thomas F.X. Noble, 'Secular Sanctity: Forging an Ethos for the Carolingian Nobility', in *Lay Intellectuals in the Carolingian World*, ed. Janet L. Nelson and Patrick Wormald (Cambridge, 2007), 16.

18. This text was edited by Madrisius in 1737 and reprinted in Paulinus of Aquileia, *Liber exhortationis, vulgo de salutaribus documentis, ad Henricum comitem seu ducem forojuliensem*, PL 99: 197–282. This text has never been translated; all translations herein are my own. The oldest manuscript containing the text dates from the ninth century: Paris BNF lat. 6649. For more description of the text and its manuscript tradition, see McKitterick, 166–68.

19. Paulinus, 5; PL 99: 200–201: 'Esto, quaeso, quamvis laicus, ad omne opus Dei promptus, pius ad pauperes et infirmos, consolator moerentium, compatiens miseriis omnium, largus in eleemosynis, memorans evangelicae viduae duo minuta (Luc. XXI, 2), et prophetam dicentem: Frange esurienti panem tuum (Isa. LII, 7), at caute praevidens discretionem eleemosynae, ita ut utrisque, danti scilicet et accipienti, solatium sit.' For Paulinus' rhetorical style, see Paul E. Prill, 'Rhetoric and Poetics in the Early Middle Ages', *Rhetorica: A Journal of the History of Rhetoric* 5, no.2 (1987): esp. 146–47.

20. Paulinus, 38; PL 99: 240: '…nec solum nobis, sed etiam omnibus laicis, ejus ex toto corde praecepta servantibus, regnum coelorum promissum est.'

21. Ibid.; PL 99: 239–40: 'Tu autem, frater mi, omnibus domesticis tuis et tibi subjectis praecipe, ut se sobrios exhibeant, et iterum propter abstinentiam in superbiam se non erigant; sed omnia temperate, juste, pie, et religiose secundum Dei adjutorium faciant: quia non solum pro nobis clericis, sed etiam pro omni genere humano, qui praedestinati sunt ad vitam aeternam, Christus sanguinem suum fudit…'

22. Ibid.; PL 99: 240–2: 'Grandis namque confusio est animabus laicorum, qui dicunt: Quid pertinet ad me libros Scripturarum legendo audire vel discere, vel etiam frequenter ad sacerdotes et ecclesias sanctorum recurrere? Dum clericus fiam, faciam ea quae oportet clericis facere…. Omnibus enim laicis, clericis, monachis aequaliter convenit fidem, spem, charitatem, humilitatem habere, et Deo ex toto corde servire: veram confessionem facere, et dignam poenitentiam ageret: quia clementissimus Dominus confugientibus ad poenitentiam ignoscit.'

23. Coon, 443–4. Cf. Gillian R. Evans, '"Interior Homo": Two Great Monastic Scholars on the Soul: St. Anselm and Ailred of Rievaulx', *Studia monastica: Commentarium ad rem monasticam investigandam* 19, no. 1 (1977): 57–73.

24. Paulinus, 3; PL 99: 199–200.

25. Ibid., 4; PL 99: 200: 'Tuo vero capiti, frater charissime, addatur gratia spiritualis scientiae…'

26. Ibid., 23; PL 99: 218–19.

27. Ibid., 22; PL 99: 216–17: 'Qui charitate plenus est, tranquillo animo et serenissimo vultu procedit; vir odio plenus ambulat iracundus.'

28. Ibid., 10; PL 99: 206: 'Proinde si aliquid in hoc saeculo possidere delectamur, Deum qui possidet omnia, expedita mente possideamus, et in eo habeamus quaecunque feliciter et sancte desideramus. Sed quoniam nemo possidet Deum, nisi qui possidetur ab eo, simus nos ipsi facti Dei possessio, et efficietur nobis possessio Deus.'

29. Jonas of Orléans, *Jonae de institione laicali libri tres*, PL 106: 121–278. This text has not been translated into English; all translations of Jonas are my own. Pierre Riché has made much of the fact that Jonas refers to the lay status

as an '*institutio*', arguing that the bishop of Orléans was bestowing upon laypeople status of their own institution within the Church; Riché, 99–100. Perhaps he was, but the title can also simply mean 'On lay instruction', or 'On lay education', For Matfrid of Orléans, see Philippe Depreux, 'Le Comte Matfrid d'Orléans (Av. 815–836)', *Bibliothèque de l'École des Chartes* 152, no. 2 (1994): 331–74.

30. Jonas' treatise makes frequent reference to both the Old and New Testaments, drawing particularly from the historical books and the letters of St. Paul, but also, liberally, from the Gospels of Matthew, John and Luke. Like Paulinus and Alcuin, Jonas quotes from the prophets, from Proverbs, and less frequently, from the Psalms. Of the Fathers, Jonas refers often to the words of Augustine, Jerome, Ambrose as well as the homilies and *Dialogues* of Gregory the Great. Jonas' text is divided into three books of 20, 29 and 20 chapters, respectively. For scholarly work on the *De institutione laicali* se especially Savigni, 71–2ff. with notes. Cf. also Étienne Delaruelle, 'Jonas d'Orléans et le Moralisme Carolingien', *Bulletin de littérature ecclésiastique* 55 (1954): 129–43, 221–8. For Jonas' life, see Raffaele Savigni, *Giona Di Orléans: Una Ecclesiologia Carolingia* (Bologna, 1989).

31. Savigni, 'Les Laïcs dans l'ecclësiologie Carolingienne: Normes Statutaires Et Idéal de "Conversion" ', 76–7. See esp. Jonas, II.6. There can be no doubt that Jonas chose to place so much emphasis on marriage in his treatise because this was precisely what Count Matfrid asked his spiritual advisor to discuss. Ibid., *praefatio*; PL 106: 122–3.

32. Jonas explained that God commanded all his earthly children to embody an extreme love of a type unknown before Christian times. All souls were expected to love both friends and enemies alike, Jonas said, for this was what separated true Christians from everyone else. Ibid., III.1; PL 106: 233.

33. Ibid., I.20; PL 106: 161: 'Lex itaque Christi non specialiter clericis, sed generaliter cunctis fidelibus observanda, est a Domino attributa.'

34. Ibid.

35. Ibid.; PL 106: 162.

36. Ibid.: ' … Licet in Evangelio quaedam sint praecepta specialia, quae solummodo contemptoribus mundi et apostolorum sectatoribus conveniant; caetera tamen cunctis fidelibus, unicuique scilicet in ordine quo se Deo deservire devovit, indissimulanter observanda censentur. Multi namque laicorum existunt, qui legem evangelicam et apostolicam sibi datam credunt, et intelligere procurant, et secundum eam vivere pro viribus invigilant. Sunt alii qui eam sibi datam credunt, hanc tamen intelligere, et secundum eam vivere detrectant.' The notion of separate 'precepts' and 'commandments' for all the faithful and for ascetics would not become doctrine until the thirteenth century (cf. Thomas Aquinas, *Summa theologica* 2.1.108.4).

37. Odo of Cluny, *Vita Geraldi*, praefatio; PL 133: 639–40; Sitwell, 296.

38. Cf. Alcuin, 34; PL 101: 636–37.

39. Odo of Cluny, *Vita Geraldi*, I.4–5.

40. E.g. ibid., I.7–8, I.11, I.18, I.20, I.32–41.

41. Ibid., I.11.

42. Ibid., I.14, I.17, I.28. For an extended study of this aspect of Gerald's character in the context of Cluny during the tenth century, see Robert Ian Moore, *The First European Revolution, c. 970–1215* (Oxford [England] and Malden,

Mass., 2000), 53–4, and Barbara H. Rosenwein, *Rhinoceros Bound: Cluny in the Tenth Century* (Philadelphia, 1982), 73–4, 77, 79–80.

43. Odo of Cluny, *Vita Geraldi*, I.11; PL 133: 649; Sitwell, 305–6: '… vitam satis honestam ducebat, et medioximum discretionis callem in neutra parte pronior agebat. Scilicet ut saeculari negotio sua munia non fraudaret, nec a religionis cultu terrena se occupatione tricaret …'

44. Ibid., I.30; PL 133: 660; Sitwell, 317: 'Merito quippe diligebatur ab omnibus, quoniam ipse cunctos diligebat.'

45. Modern cultural theorists have demonstrated how hybrids point far more toward conflict than toward the sort of harmonious melding of forms that they seem initially to display. For the most recent bibliography concerning hybridity in medieval literature, see Jeffrey Jerome Cohen, *Hybridity, Identity, and Monstrosity in Medieval Britain: On Difficult Middles* (New York, 2006). It must be said, however, that the self-conscious theorization and discussion of hybrid identities has entered into medieval scholarship only tentatively, rarely concerning the period before the year 1000, at least for the continent, and never, to my knowledge, discussing monastic/secular hybridity.

46. Odo of Cluny, *Vita Geraldi*, I.9, I.10. See Nelson, 'Monks, Secular Men and Masculinity, c. 900'; and more recently, Jacqueline Murray, ' "The law of sin that is in my members": The Problem of Male Embodiment', in *Gender and Holiness: Men, Women, and Saints in Late Medieval Europe*, ed. Samantha Riches and Sarah Salih (London and New York, 2002), 13–17.

47. Odo of Cluny, *Vita Geraldi*, I.19, I.26, II.18; PL 133: 657–8; Sitwell, 315: 'Quis alius praeter Geraldum hoc ita faceret? Certe mihi videtur, quod id magis admiratione dignum sit, quam si furem rigescere in saxi duritiam fecisset.'

48. Ibid., I.8.

49. Ibid., PL 133: 646; Sitwell, 302.

50. Ibid., PL 133: 647; Sitwell, 302.

51. Ibid.: 'Nemo sane moveatur, quod homo justus usum praeliandi, qui incongruus religioni videtur, aliquando habuerit. Quisquis ille est, si justa lance causam discreverit, ne in hac quidem parte gloriam Geraldi probabit obfuscandam.'

52. Ibid., II.2, II.3.

53. Ibid., II.4; PL 133: 672–3; Sitwell, 329.

54. Ibid., PL 133: 673.

55. Ibid., II.6; PL 133: 674.

56. Ibid., I.24; PL 133: 656: 'Erga ipsos vero qui juris ejus erant, tam beneficus, tamque erat pacificus ut mirum considerantibus esset. Nam frequenter improperabant et quod mollis esset et timidus, qui se laedi ab infimis personis, tanquam impotens, permisisset.' For softness as a negative feminine quality, particularly in the monastic context, see Coon, 436–7.

57. Ibid., II.8; PL 133: 675; Sitwell, 331.

58. For a very recent discussion of such phenomena, see Katherine Allen Smith, 'Saints in Shining Armor: Martial Asceticism and Masculine Models', *Speculum* 83, no. 3 (2008): 572–602.

59. In this I agree with Ruth Mazo Karras' recent argument for 'a fluidity of meaning within the binary categories of masculine and feminine' rather than a spectrum of genders between masculine and feminine poles, or any

sort of 'third gender' category represented by chastity; Ruth Mazo Karras, 'Thomas Aquinas's Chastity Belt: Clerical Masculinity in Medieval Europe', in *Gender and Christianity in Medieval Europe: New Perspectives*, ed. Lisa M. Bitel and Felice Lifshitz (Philadelphia, 2008), 53ff. For the counter-argument, see in the same volume, Jacqueline Murray, 'One Flesh, Two Sexes, Three Genders?', 34–51.

60. Caroline Walker Bynum, *Holy Feast and Holy Fast: The Religious Significance of Food to Medieval Women* (Berkeley, 1987), 1.

3

The Warrior *Habitus*: Militant Masculinity and Monasticism in the Henrician Reform Movement

Scott Wells

The German King and Emperor Henry II (r. 1002–1024) actively promoted a specifically Benedictine renewal of the *Reichsklöster* (monasteries under direct royal protection and patronage) in his kingdom. As a reformer, Henry II reflected a widespread phenomenon of lord-initiated monastic reform in the period from ca. 900 through the mid-eleventh centuries.[1] Alongside several dukes, counts, bishops and some of his fellow kings both on the continent and in Anglo-Saxon England, Henry II can be seen as the focus of a distinct monastic reform initiative defined as much by his top-down authority as by the bottom-up collaborative contributions made by reforming monks.[2] Henry II, however, championed an agenda with distinct characteristics, including a particularly strong emphasis on a masculinization of the cenobitical *habitus* – of the environment constituted by and in turn constituting the thoughts and practices of the claustral community. Henry II promoted the image of monasticism as a collective enterprise, at once hierarchical and communal in its structure and demands, requiring both strong leaders and effective foot-soldiers, shaped by the camaraderie of a shared struggle on behalf of God – truly a spiritual *res militaris*, driven by the need to improve the preparedness of monks (but also nuns) for this warrior *habitus*.[3]

Henry II's initiative to masculinize the monastic order in Germany, to ensure that the *milites* and *superiores* of the *Reichsklöster* conformed more exactly to the complete discipline of the Rule in the service of God, had a powerful contemporary impact and left numerous traces in the surviving record. The militant ideology of the project was articulated in Henry II's official grants of privileges and properties to Benedictine houses in Germany, in the contemporary *vitae* of two abbots who numbered among Henry II's chief monastic associates, and even in a poem of support drafted by Abbot Odilo of Cluny – sources which imagine the ideal cloistered life in emphatically bellicose terms. On the other hand, several contemporary monastic chroniclers voiced explicit opposition to both the masculinist principles and exclusionary implementation of the monarch's reform agenda. These perspectives opposed the disruptive, discriminatory implications of Henry II's call for greater militarization of the monastic life with what they constructed as an older and superior ideal of harmonious, egalitarian unity focused on a collective oneness encompassing past and present, weak and strong and monks and abbot. The hierarchies imposed by the warrior *habitus*, in contrast, called for the subordination of inferiors to superiors, with 'superior' defined explicitly in terms of the commanding warrior male; those who could not successfully follow or subordinate themselves to the orders of the leader needed to be expelled from the community to preserve the discipline of the whole. Together, these sources provide contrasting perspectives on the renewed emphasis on the militant manliness of Benedictine monasticism in early eleventh-century Germany.

In examining how masculinity and monasticism were associated by Henry II and his reform circle, and how resistant voices disputed that linkage, the present analysis reinforces some of the theoretical observations made about monastic and clerical masculinity over the past 15 years while challenging others. Much of this scholarship has concentrated on medieval definitions of masculinity associated with sexual potency. In a foundational essay on 'Being Male in the Middle Ages', Vern L. Bullough cites anthropologist David D. Gilmore's observation that societies tend to define manhood around 'a triad: impregnating women, protecting dependents, and serving as a provider to one's family', but then describes medieval manhood as almost exclusively defined by sexual performance, rather neglecting the political and economic performance also highlighted in Gilmore's definition.[4]

Celibacy certainly functioned as a key category in distinguishing monks from secular priests in the early Middle Ages. As Jacqueline Murray observes, 'priests tended to remain active in society, caring for their flocks, earning a living and frequently enjoying marriage and family.'[5] In this pre-Gregorian world, chastity was a specifically monastic discourse, 'directed as much against the behavior of married or sexually active secular clergy as it was against the temptations of lustful women'.[6] It is also true that the regular clergy were in this sense 'monasticized' in the later Middle Ages. Starting with the eleventh-century Gregorian Reform, and accelerating in the twelfth and thirteenth centuries, priests were expected to become more like monks, which meant specifically to become celibate.[7] This emphasis on the ways in which sexuality defined clerical masculinity has produced stimulating if sometimes overly reductive and contradictory arguments, from arguing that celibacy turned priests into misogynists who associated perfect maleness with sexual purity to claims that the prohibition of sexual potency had the effect of 'emasculating' the clergy and making them strongly sympathetic to religious and lay women.[8]

However, for understanding the warrior masculinity that Henry II and his collaborators strove to cultivate in the German *Reichsklöster*, those studies of clerical masculinity which prioritize the association of manliness with demands of leadership, discipline and responsibility to protect the welfare of *familia* and dependents – rather than the demands of sexual penetration – prove much more useful. Thus, Maureen Miller's important essay on masculinity and reform during the Gregorian era uses a close reading of the *vitae* of the saintly bishops Ulrich of Augsburg (d. 973) and Ubaldus of Gubbio (d. 1160) to demonstrate that the clergy of the eleventh and twelfth century were not embracing or transcending gender, but were rather engaged in 'a competition between clerical and lay males over degrees of masculinity'.[9] This clerical culture of reform defined the masculinity of lay men in fundamentally sexual terms – through bodily, violating interaction with women and therefore (within the logic of this discourse) as violent, quarrelsome and destructive.[10] Clerical men contrasted this polluted and sexual form of lay masculinity with their own higher masculinity defined in terms of leadership and responsibility to secure 'the chaste communal life of the clergy and the defense of the church against the depredations of the laity'.[11] The ideal type of the masculine cleric was 'the lone manly bishop who single-handedly brings peace and prosperity to his city'; or in other words, 'a *pater*, an *athleta Dei*, and a *heros*, all words with strongly masculine connotations, the latter two … usually associated with superlative physical, military, political, and moral fortitude.'[12] The world view 'opens up a moral chasm between clerical males and lay males' in which lay men as whole define bad, disorderly, sexual masculinity and clerical men alone define good, powerful, protective masculinity.[13] While still highlighting an association of masculinity with sexuality, and the misogyny of a clerical discourse that equated bodily contact with women (even in the form of breast milk) with a lesser form of masculinity,[14] Miller's argument shifts the balance away from genital maleness toward paternal, athletic, political maleness in a way that better reflects and dissects the nuances in Gregorian reform discourses about celibate clerical authority.

Janet L. Nelson, Emma Petit and Dawn Marie Hayes have similarly focused on the ways in which medieval gender discourses in Carolingian, Anglo-Saxon and Capetian monastic and royal contexts associated true masculinity with the warrior virtues of just obedience and wise command, rather than the flawed manliness of raw physical prowess and sexual virility. They have, however, also highlighted how this discourse of masculinity could bring kings, nobles, bishops and monks together in a shared militant manliness rather than pitting clergy against laity as two different kinds of men. Indeed, this ideal could also incorporate women into its vision of manliness, at least as fellow warriors in spiritual combat if not in physical potency, aggression or heroism.[15] Henry II's reform was built upon a similar ethos, calling for a partnership between king and monks to 'virilize' the cloister, ensuring its increased masculinization through renewed adherence to the militant virtues of discipline, leadership and service. Furthermore, while it is revealing that the movement assigned public leadership roles exclusively to men, the warrior *habitus* required by Henry II's interpretation of the Benedictine Rule nevertheless applied equally and without distinction to both monks and nuns. Indeed, as will be shown below, when contemporaries promoted or challenged Henry II's reform efforts, they were not primarily debating definitions or degrees of virility, or contrasting masculinity with femininity, but arguing over whether a militant manliness constituted the proper measure by which to evaluate standards of leadership, service, discipline and communal belonging within the German *Reichsklöster*.

I   The implementation and ideology of Henry II's monastic reform initiative

According to Hartmut Hoffmann, no German king dedicated himself more energetically to claustral reform than Henry II.[16] The spiritual goal of this renewal was to improve adherence to the dictates of the Rule among the monastic communities in Henry II's German realm by allowing the king to exercise his divinely mandated authority over and responsibility for correct Benedictine discipline in the kingdom's cloisters. The king's reform initiative certainly also incorporated the economic (and political) goal of improving the administration of monastic property to improve the provision therefrom of *servitium regis* – that is, goods and services for the support of the itinerant court and royal army.

But as articulated and enacted by Henry II and his supporters, the religious agenda to centralize, masculinize and militarize the supervision of claustral discipline was paramount.

Henry articulated the militant ideology behind this reform agenda in the texts of his grants of privileges and properties to royal monasteries, In these charters, he formulated several original expressions of his political and religious convictions.[17] Within royal charters (or *diplomata*), these two sections served to contextualize and explicate why the privilege or property was being granted. In a charter of 1013/1014 perpetually granting possession and control of the monastery of Schwarzach to Bishop Werner of Strasbourg and his successors, Henry II incorporated the following passage:

> As the shape of the human body has been arranged according to a rational structure by omnipotent God, so that the lesser limbs are made subordinate to the head and are ruled by it as though by a military commander, we consider it not discordant, in imitation of this model, to place certain lesser churches in our kingdom under the greater ones; and we have determined that this in no way obstructs the will of the king of kings, who knew how to arrange celestial and earthly sovereignties in such a wondrous order.[18]

Although this vision reflects what Stefan Weinfurter has demonstrated to be Henry II's broader project of political centralization in his kingdom, it is expressed here in a specifically monastic context: it is the king's responsibility to see that all the churches (including monastic ones like Schwarzach) are well-ruled.[19] Those that lacked the proper discipline (and/or economic foundation) to function as a true *Reichskloster* must be subordinated either to the oversight of a reform abbot or (as here) bishop appointed by the king. In fact, according to Hubertus Seibert's calculation, Henry II stripped 21 *Reichsklöster* of their *libertas* during his reign, placing 17 under the control of bishops.[20] Henry's reform initiative also involved identifying particularly effective clerical leaders (such as the abbots Godehard of Niederaltaich, Immo of Gorze and Poppo of Stavelot, or the bishops Werner of Strasbourg and Meinwerk of Paderborn) and assigning them several monasteries to reform. These implementers of reform would then preside over multiple cloisters until they were ready to pass control of one or more of them on to their chosen students trained for that purpose.

In another charter, granting the monastery of Abdinghof to the bishopric of Paderborn in 1017, ruling is presented as a specifically masculine activity and being ruled as a feminine one, highlighting an association of manliness not with sexual prowess but with a contemplative vigilance that prepares and qualifies one for rule:

> The universal wisdom of all Christ's faithful knows that there are those among humankind who possess the strength to rule like a male, and those who like a female are to be ruled. In reflecting upon this as a contemplation of divine love – being in interior things like the sentinels of the night and in exterior things like those who are sleeping – and at the intervening of the most cherishable women, our wife and the august empress Cunigunde, we grant ownership of that farming estate to the new church which the revered Bishop Meinwerk constructed just to the west of the city of Paderborn and completed the dedication to holy Mary and all the saints…so that the *fratres* serving God with charitable concord under the rule of St. Benedict at that same monastery may hold and possess it with a most powerful proprietary right.[21]

In the contrast of feminine subservience with strong, masculine leadership, Henry II's contemplation is associated with the latter, but so also are Cunigunde's effective intervention, Bishop Meinwerk's ecclesiastical patronage and (of particular note here) the monks serving God beneath the Benedictine rule (*fratres deo sub regula sancti Benedicti servientes*). The language of this charter provides an instructive example of what does, and does not, seem masculine when that quality is exhibited not through sexual prowess or corporeal strength, but instead through those qualities – including a properly disciplined humility in the service of God – that make one qualified to lead.

In many of his *diplomata* for cloisters, Henry used the phrase *fratres* or *sanctimoniales sub regula sancti Benedicti deo militantes*: monks battling for God according to the rule of Saint Benedict. This was not an entirely new formula in the royal/imperial chancery. The description of monks as *fratres* or *monachi militantes* had been used sporadically in the charters of Henry II's tenth-century royal predecessors.[22] Charters of Otto I to Einsiedeln and St. Remi at Reims had referred to monks as battling 'according to the discipline of the rule' (*sub regulari disciplina militantibus*), phrasing also adopted by Otto II and Otto III in charters to these same communities.[23] For the monks and nuns who received his privileges at the cloisters of Petershausen, Waldkirch, Luisberg and Seeon, Otto III proposed another formulation: 'those who serve God according to the rule of Saint Benedict' (*deo sub regula sancti Benedicti servientium*).[24]

Only once, however, prior to Henry II is the phrase *et monachis sub norma sancti Benedicti deo ibi militantibus* used, combining the concepts of living under the rule of Benedict with the notion of performing battle or service for God: in Otto III's November 4, 994 charter for the monks of Petershausen.[25]

A phrase used in a single charter by Otto III took on heightened importance under Henry II, who used it repeatedly. The new king seems to have found this particular formula especially apt for expressing his vision of what the monastic life entailed, and why the monarch was obligated to nourish it. Variants appear for instance in charters for Selz ('to the monks who battle there according to the rule of the blessed Benedict'[26]), Epinal ('for the sacred virgins who according to the norms of the sainted father Benedict battle for God with virility and serve God with humility'[27]), Einsiedeln ('for the servants in that cell battling according to the rule of Saint Benedict on God's behalf through the coming times'[28]), Nienburg ('for the augmenting of that place and of the *fratres* there who battle for God beneath the rule of Saint Benedict'[29]) and Kühbach ('that monastery will be returned without encumbrance to the power of the abbess and the holy nuns who there battle for God as comrades-in-arms beneath the rule of Saint Benedict'[30]). In its widespread use, this rhetoric is uniquely Henrician, and Henry II's immediate adoption of this metaphor into his monastic charters must have had a profound impact. It certainly made a strong statement of how a new specific, vision of monastic life – as conjointly Benedictine and militant – was now being promoted under the monarch's tutelage.

Interestingly, after ca. 1017 these expressions of claustral militancy are replaced in Henry II's charters with formulae like 'the nuns who in our kingdom justly live according to the rule of Saint Benedict' or 'for the utility of the brothers and the church where they serve God beneath the rule of Saint Benedict'.[31] In grants both to communities of women (e.g. Kaufungen, Göss and Dietkirchen) and of men (e.g. Abdinghof, Tegernsee, Prüm, St. Emmeram and Heiligenberg), Henry II now regarded his Benedictines as *sub regula sancti Benedicti servientes* rather than *militantes*.[32] Yet Henry's reform ethos had always embraced both aspects of positive masculine leadership – humble service as well as virile battling – a related pair described by Janet L. Nelson as the Solomonic and Davidic models.[33] Even in 1003, Henry II described the nuns of Epinal as embodying both qualities.[34] Humility, far from being a contrasting feminine virtue, was an integral part of an effectively militant masculinity – as borne out in the contemporaneous *vitae* of two of Henry II's chief reform abbots, Godehard of Niederaltaich and Poppo of

Stavelot – composed not to promote the saintly cults of these men, but instead to justify the Henrician reform initiative and their association with it.

II Expressions of monastic militancy in the reform circle of Henry II

The *Vita prior* of Godehard of Niederaltaich was written in 1035 by his disciple Wolfher.[35] Godehard was still alive, and had been elevated to the bishopric of Hildesheim in 1022, where Wolfher was one of the canons. The *vita* was composed, however, at the request of Ratmund, Godehard's nephew, who had succeeded him as abbot of Niederaltaich, and thus in its description of Godehard's monastic career is attentive to a primarily monastic audience.[36] Wolfher begins by describing how, at the time of Godehard's birth, Niederaltaich had devolved into a canonry under the dominion of Archbishop Frederick of Salzburg. Godehard's father, ennobled by virtue if not by birth, had descended from a line of dependents of the monastery. Though a layman, the archbishop appointed him prior of the cloister. His son received his formative education at the same *coenobium* from the priest Oudalgisus, 'renowned for his worthy discourses (*conversatio*) of canonical instruction and the most celebrated in all that region for the authority of his scholarly *studium*'.[37] Eventually, by the collective election of the *fratres*, because they cherished him in a unanimous love, Godehard was himself ordained prior of the community. Duke Henry (Henry II's father), however, decided in consultation with the bishops of Passau and Regensburg to convert the *coenobium* to a better understanding of the monastic *ordo*. He committed Niederaltaich to the Swabian Erchanbert, a man of just *conversatio*, in order that the *regula sancti Benedicti* be instituted at that place. The members of the community were given the choice either to become monks or, if they wished, to leave. All chose to depart except Godehard who, although he was young in age, excelled all others in intelligence and wisdom; the love the other *fratres* felt for Godehard is thus, in Wolfher's Henrician narrative, not sufficient to ensure their adherence to strict monastic discipline under the Rule.

Godehard alone from the old community remained to train as a Benedictine monk under Erchanbert's leadership, and so far excelled in this role that when the future Henry II became Duke of Bavaria he sought to remove Erchanbert and make Godehard abbot instead. Godehard declared at length both his unwillingness to accept the overthrow of his teacher and his own unworthiness to assume the abbacy. Of course in the ideology of Henrician reform this humble obedience, so wisely and eloquently expressed, demonstrated his qualities for monastic leadership even further. He then sought refuge among the *fratres* who battled for God at the abbey of St. Emmeram under their wise abbot (Ramwold). Finally, two years after Erchanbert's deposition, and mostly in response to the tearful clamor of the congregation of his (Niederaltaich) *fratres*, joined by the *familia* of monastery, he relented and took up the abbatial office.

While Wolfher thus uses Godehard's progress from childhood to abbot to define the model of monastic leadership exemplified by the reformer, that model is most clearly and distinctly on display in the *vita*'s descriptions of Godehard's interventions at Hersfeld (1005) and Tegernsee (1001), which are narrated in reverse chronological order. In the fourth year of the reign of Henry II, Abbot Bernhar of Hersfeld 'praiseworthy in his worldly dignity' (and thus by implication *not* in spiritual leadership) died,

> whom the entire unity of the catholic faith, with the mediation of King Henry, appointed the Lord Godehard to replace with the purpose that he should lead back to the right way of life the monks there who indeed were living not according to the rule or even canonically, but ostentatiously and effeminately (*enervate*).[38]

Since it is written, 'Begin at my sanctuary' (Ezech. 9:6), he offered the *fratres* two choices:

> either to conform to their rule, namely that of St. Benedict, according to their intellectual capacity and physical capabilities; or, if they preferred, they could depart unhindered through the open gates.[39]

In Erchanbert's staggering figures, no fewer than 50 left, with no more than two or three remaining behind with the abbot. However, two or three more returned after a month, an additional three or four after a year, and indeed in the seven years Godehard retained control of the convent all but two or three of the monks had returned. Godehard's struggle to restore the monks of Hersfeld, undertaken at the commission of Henry II (who in turn was acting on behalf of the entire concord of the faith), thus ends in an almost total victory thanks to the combination of strong discipline (in imposing the rule) and gentle humility (in allowing the dissatisfied to leave and, when they desired to return) demonstrated in his leadership.

Turning to Tegernsee, Wolfher writes how in the region of Bavaria there was a certain *coenobium* of monks lacking pastoral care which was similarly commended to Godehard so that he could demonstrate the path of true religion to the monks residing there. By dividing unlawful practices from lawful ones, he was able to reform those of their customs by which that place now everywhere excels. Indeed, continues the *vita*, Godehard was able to implement this transformation with great rapidity, 'because divine favor – going before, accompanying, and following after – assisted him in all things'.[40] Later, following his advice, the seats of governance of the two monasteries Hersfeld and Tegernsee were commended to two of his monks – Arnold and Burchard – whom he himself had nourished, and whose life and *mores* he therefore knew to be worthy, and whom indeed he had previously appointed as leaders second only to himself at the *coenobium* of Hersfeld (by which Wolfher appears to be describing a form of abbatial of apprenticeship).[41] Arnold assumed control of Hersfeld in 1012 and Burchard of Tegernsee in 1013, but the latter was actually the third abbot sent by Godehard to follow him at the Bavarian cloister, following Eberhard (1002–1003) and Peringer (1003–1013). Wolfher, by narrating Godehard's reform of Tegernsee *after* that of Hersfeld, and by silently passing over the existence of Eberhard and Peringer, is thus able to construct a narrative representation of Godehard's success as a claustral leader which culminates in an orderly transfer of authority of two reformed communities to two well-prepared adjutants. Godehard's service for God beneath the Benedictine rule thus receives an extra burnish in Wolfher's portrait of a man whose mission and militant masculinity (displaying at once the mastery of a commander and the discipline of a foot soldier) are expressed in terms and concepts closely mirroring those in Henry II's *diplomata*.

Poppo of Stavelot, as described in the *vita* by his nephew and monastic disciple Everhelm, abbot of Hautmont, likewise demonstrates a combination of humble obedience and disciplined lordship that effectively protected and enhanced the regular life of his *fratres*.[42] Poppo's father Tizekinus was a high-status warrior who died in battle, leaving his illustrious wife Adalwif a widow and his son an orphan; but Poppo therefore had cause to humbly rejoice in the song of Psalm 26:10, 'My father has abandoned me; God, however, has taken me up.'[43] The remainder of Poppo's youth continues this theme. At first following in the footsteps of his earthly father, he inanely applied himself to training in secular weapons, and then strove greedily along with his warrior-companions after an empty glory.[44] However, after actually experiencing the *res militaris*, while he prevailed in skill and audacity, nevertheless he saw the futility, danger and likelihood of defeat. Having placed himself at the entryway to the world, he now wished to draw back his foot.[45] Though circumstances required that he remain in military service, he began to live like a *miles Dei*:

> although by outward appearance he was untonsured under lay garb, and still a fighting warrior, and indeed pretended to bear arms in the *militia* under the command of Theoderic, one of the most illustrious men of that time, under this military service he put on the armature of priestly religion, and displayed the insignia of monastic devotion in his works.[46]

Eventually, following a pilgrimage to Jerusalem with two of his fellow soldiers, whom he then persuaded to become monks, Poppo was able with the support of his commander Theoderic to enter the monastery of St. Theodoric at Maroilles under the leadership of Abbot Eilbert, again reinforcing the parallels between military service in the world and in the monastery.

Poppo soon found a new and superior monastic leader in Abbot Richard of Saint-Vanne, continuing the theme that a successful monk, like a soldier, needs a proper commander under whom to serve. Richard met Poppo during a visit to Maroilles, and was so taken with his virtues that he humbly beseeched Eilbert to transfer the young monk to him. At Saint-Vanne, Poppo demonstrated perfect obedience to Richard in all things, and 'being found in all things obedient, faithful, and patient from the valuing of servitude, he soon was held [by Richard] in the highest love among all the other sons adopted by him.'[47] Everhelm goes on to describe how Baldwin IV, Margrave of Flanders, summoned Richard to take over the administration of Saint-Vaast from the ejected Abbot Folrad. Richard took on this charge as ordered, and sent Poppo to preside over that monastery in his place. Poppo succeeded in winning the consent of all the *fratres* there, and also restored the economic foundation of the community by working to have a benefice or *res monasterii* unjustly occupied by some local *milites* restored to the *fratres* with the assistance of the margrave. Despite his success there, Poppo subsequently left Saint-Vaast at Richard's command and returned to 'the originary place of his [monastic] *conversatio*', and with patience and obedience undertook as imposed on him by the abbot the administration of all the things that needed correcting in the monastery.[48] Seeing that he confronted every burden with patience and humility, and that no labor was too great for his obedience, he appointed him for a second time to command a monastery, in this instance Wasloi.[49] On his arrival, Poppo discovered that the buildings of the monastery (both the liturgical structures and those devoted to agriculture and manufacture) had decayed and were in any case too small. He had everything elegantly rebuilt from the bottom up and even gave the monastery a new name, Beaulieu or 'the pretty place'.[50]

This is the reformer and leader, as recorded by Everhelm, whom Henry II summoned to lead and renew the *Reichskloster* of Stavelot in 1020 following the death of Abbot Bertram. The man remembered by God, promoted to the direction (*regimen*) of the monastery, began to displease those residents of the place whose customs (*mores*) he found unsuitable and tried to correct by his example of acting well (*bene agendo*). The strength of the animosity between those who were originally resident in the place and the *fratres* Poppo brought with him began to grow so great, sometimes in manifest hatreds and sometimes in clandestine plots. A wrath raged furiously among the laity who were brought into the conspiracy, 'so that on a certain day they erupted into the monastic cloisters, prompted by [those who hated Poppo's corrections], and chased the lambs of Christ through the lord's enclosure with drawn swords'.[51] Miraculously, those who seemed to be cut down by the swords flashing among them were unharmed, kept from wounds by not a human but a divine shield, just as the bush which seemed to Moses to be on fire but suffered no harm.

As he brings his chapter on the reform of Stavelot to a climax and conclusion, Everhelm continues the association of Poppo with Moses as a victorious leader in spiritual battle:

> The true report of this great evil reached the ears of the blessed man Poppo, then staying at Malmédy, and caused him to clothe himself in a spirit of contrition and grief on account of the accursed appearance of things, especially in receiving ambassadors from their party, when he prostrated himself before them. Then those who still stood firm began to feel anxiety about remaining [at Stavelot]. There were, besides, two men of the secular soldiery (*militiae saecularis*) who came to know of what had happened, of whom one chanced to be name Adalbert, the other Boso. With the same great speed with which that phalanx [of invaders] had attacked the soldiers of Christ (*tyrones Christi*), these [two men] proceeded to go there and, through God rather than with the application of their own virtues, they alone expelled, took captive, and scattered all that multitude; and in all this the thing was done according to an oath. In the merits

of these two men, aided by the expelling hands of the man of God, that one fought who saves not through thousands or forces of chariots and weapons, but who with one puts a thousand to flight, with two, ten thousand. [Deut. 32:30] Thereafter, it was enacted through divine punishment, that not one of those who had been in that conspiracy of the iniquitous survived the year. Thus in everything that concerned the man of God confusion gave way to grace, and love brought forth an equal harmony of all in him.[52]

As was prophesied to Moses, God smites the enemies of his people through the agency of one or two, and their destruction is total. Adalbert and Boso aid Poppo like another Joshua, and Poppo in turn assists them with his expelling hands as Moses used his upraised hands to assist Joshua in expelling the Amalekites (Exod. 17:8–13). Rather than force, Poppo relies on his awe-inspiring humility and trust in God to resolve the crisis, and this display of monastic masculinity in the Henrician mode allows Poppo to prevail entirely, and achieve a perfection of discipline in Stavelot under his leadership. As Everhelm goes on to observe, Poppo had to overcome similar resistance from the monks at St. Maximin at Trier when he was appointed to succeed Abbot Hierich there 2 years later. In this instance, opposition came specifically 'from those whose way [of living] had strayed far from the monastic regulations and observations'.[53] In this case the monks supposedly used magic (*praestigium*) in order to dispose of their unwanted abbot by poison; but once Poppo's viscera had miraculously withstood this diabolical act and his enemies had been disposed of by God (again within a year), he was able to implement reform. Once more the display of a specific kind of masculine strength, in this instance his body literally withstanding poison with divine assistance, allowed him to cow opposition and impose the disciple of the rule on a formerly recalcitrant community.

The perspective conveyed in these *vitae* of Godehard and Poppo is that of the central participants in Henry II's reform movement, those closest to (and most dependent upon) imperial authority. One of the clearest statements of the ethos behind this initiative, however, comes from the pen of one of its outside supporters, Abbot Odilo of Cluny. The Henrician reform was not directly affiliated with Cluny, and Cluny was not directly subordinate to the German emperor. Nevertheless, the relationship between Odilo and Henry II was close, and Odilo's influence on Henry II's monastic ideals has frequently been noted.[54] Henry II granted a charter to Odilo for the Cluniac cloister of Peterlingen in 1004, and Odilo successfully interceded with the king for the granting of privileges to at least three Italian cloisters.[55] They met in person multiple times between 1004 and 1014, and perhaps as late as 1022.[56] More strikingly, several contemporary sources (including Jotsald's *Vita Odilonis* and Rodulphus Glaber's *Historiae*) record how Henry II sent his imperial crown and insignia to Cluny, suggesting a genuine partnership of mutual patronage and protection between cloister and monarch.[57] Similarly, Henry II enjoyed the unique privilege (not shared by his predecessor or successor on the imperial throne) of entry into Cluny's necrology as an *amicus* of the monastery's *societas et fraternitas*.[58] Odilo also sent at least two manuscripts to Henry II, accompanied by dedicatory poems, to assist that monarch with his monastic reform in general and the foundation of the cloister of St. Michael at Bamberg in particular.[59] Of particular noteworthiness in the present context was a now-lost manuscript of Cassian's *Institutes* and *Collations*, the texts that provided the monastics of Latin Christendom with extensive instruction in disciplining the body, emotions and thoughts based on the examples of the Egyptian desert fathers. The dedicatory poem survives in a later medieval copy, and demonstrates how Odilo and Henry II shared a common understanding of the monastic discipline as a form of military exercise. Its opening lines read as follows:

> The Romulan [i.e. Roman] realm obeys the imperial command
>   of Caesar
> And both the Teutonic and the Latin soldier serve him
> With an equal striving under material weapons
> So that our *res publica* may be expanded through their
>   accustomed firmness.
> Yet I, a weak man who does not use such weapons,
> Wish devotedly to offer to my Augustus at all times
> Through perpetual prayers accompanied by our (Odilo's) *fratres*
> Life, strength, and peace together with health;
> And so that, when wars are waged against the enemies of his peace,
> I who am always holding leisure dear may not seem to provide less
>   auxiliary support
> I offer from our spiritual weapons (*armis*), which *are* also weapons,
> A gift which is small but apt for the minds of they
> Whom that sword-belt of the sacred militia adorns,
> The disciples whom Christ commands to gird their loins.[60]

Monks, in this imagery, serve Christ and the emperor as *milites*, and are just as essential to imperial prosperity as his warriors who wield weapons of iron and steel; and the prayers of the *militia sacra* are just as powerful

as the sword. The remainder of the poem goes on to praise several
of the ancient Egyptian monks as exemplary spiritual warriors in this
mold, prototypes for the militantly masculine monks being cultivated
in Henry II's realm for service under Benedict's Rule.

III   Voices of resistance

The Henrician reform met with considerable resistance in not a few
of the *Reichsklöster*, as Wolfher and Everhelm both acknowledge. These
opponents viewed the reformers' ideologically motivated quest to militarize claustral discipline as an uncalled-for disruption and alteration of
customs and practices which had for centuries maintained the monasteries at a high standard of spiritual devotion, intellectual achievement
and confraternal love. Rather than charters and biographies, the preferred medium for expressing resistance to the Henrician reform was the
chronicle. In contrast to texts like Godehard's and Poppo's *vitae*, which
concentrated on recent and current presence of the individual reform
hero, chronicles extended their narratives back for centuries, allowing
for the articulation of a nostalgic vision which held up the memory of
the collective past as a model for emulation. Three main examples will
be considered here: Thietmar of Merseburg's *Chronicon*, the *Annales* of
Quedlinburg and Hermann of Reichenau's *Chronicon*.

Thietmar of Merseburg, a bishop who had received a monastic
education at the cloister of Berge, voiced strong opposition to
Henry II's monastic reform in his contemporaneous *Chronicon*.[61] Thietmar recounts three examples of Henry II's monastic interventions:
Fulda, Corvey and Memleben. In Book VI, chapter 91, Thietmar records,
'In those days [April/May 1013], Abbot Branthog of Fulda was deposed
and succeeded by the *conversus*, Poppo, then pastor of Lorsch. Thereafter,
as the brethren (*confratres*) scattered far and wide – what an affliction!
(*prochdolor*) – the monastery's previous *status* was altered.'[62] Poppo of
Lorsch's status as a *conversus* – an adult convert to the monastic life, but
here with the implications (a) that this makes him no true monk and (b)
that he is someone obsessed with change, in this case to the detriment
of Fulda. The harmonious *status* (condition/order/structure of being)
which the monastery had possessed has been destroyed, as represented
by and embodied in the shattering dispersal of the former *confratres*.
At Hersfeld, such a dispersal (at least according to Wolfher) allowed
Godehard to demonstrate the superiority of his reforms, thus prompting
the prodigal to return; for Thietmar, the dispersal of the Fulda brother-
hood was permanent, and highlighted the travesty of Poppo of Lorsch's

conversio. The reader of Thietmar's *Chronicle* could also contrast this destruction of Fulda (which, although Thietmar avoids directly saying so, occurred at the command of the emperor) with Otto I's restoration of the church of Fulda following a fire in Book II, chapter 42: 'The church of Fulda, which had – *pro dolor!* – burned down was restored (*renovata est*) under this emperor.'[63]

An even stronger textual resonance linking (and hence contrasting) past and present can be found in the case of Corvey. Henry II's 1015 reform of the convent is described in Book VII, chapter 13:

> ...the emperor arrived at Imbshausen on the vigil of Pentecost and observed that holy solemnity with Bishop Meinwerk. There too, Abbot Wal of Corvey was deposed, having already been suspended from his office. Without the brothers' consent (*sine fratrum consensu*), Druthmer [Druhtmar], a monk of the monastery of Lorsch, was installed in his place. During the same week, he went to his new seat. Except for nine brothers, the entire congregation departed in tears. Just as Abbot Liudolf [965–983] predicted, they had been forced to abandon this virtually empty place.[64]

The pattern of intrusion against the collective will of the monks, followed by a shattering of that former harmonious unity in a scattering of dispersed monks, is identical with Thietmar's account of the Fulda incident; but here both the role of the emperor *and* the call to the reader to compare this event with the chronicle's earlier narratives about Corvey are explicit. Although the reader will seek in vain for an explicit statement of the prediction Thietmar alludes to here, the *Chronicon* incorporates four historical narratives exemplifying Liudolf's saintly and visionary qualities. In the first two of these, the blessed abbot receives mystical visits from the souls of the recently deceased archbishops of Mainz and Cologne so that he could announce their demise to the *confratres* (or summon them to sing a requiem mass at the dead prelate's command).[65] Liudolf's third vision involves the soul of a deceased layman, the Thuringian count Gero. This man had been executed by imperial decree following an ambiguous judicial duel, despite the fact that Duke Otto of Bavaria and Count Berthold of Schweinfurt 'rebuked the emperor for allowing such a great man to be condemned on such a petty charge'.[66] Following his account of this incident, Thietmar continues:

> Here I may briefly relate the admirable service of Abbot Liudolf of Corvey, to whom God has deigned to reveal much while the

venerable father was laboring at his many vigils and fasts. At dawn on the day of the duel as he was celebrating the mass according to his custom, with humility and the fear of God, he saw Count Gero's head floating above the altar. After he had finished the first mass, he sang a mass for the dead and, after removing his vestments, silently left the church. Then, summoning the brothers (*congregatis fratribus*), he told them of Gero's death and humbly asked that they offer prayers for him in common. The beheading took place on the same day, at sunset.[67]

Again, using his divine visionary gift, Liuthar the good shepherd collects all the monks together to perform their collective work of prayer, interceding for the soul of a man unjustly condemned. In a miniature portrait of a model monastic community as imagined by an opponent of the Henrician reform, the abbot uses love and humility to request rather than the authority of office to command, and all are brought together in a shared work of mercy.

In Book IV, chapter 70, introducing the final appearance of Liudolf of Corvey in the chronicle, Thietmar again explicitly connects past to present. In those days, one of Liudolf's young monks happened to carry a container of martyrs' relics carelessly (*incuriose*), as a result of which the martyrs struck him down dead, since one who neglects to serve God's saints spiritually (*spiritualiter*) must die in the flesh (*carnaliter*). The martyrs then went to accost the abbot at the doors of the church, to announce the wayward monk's death and declare that the deed would not remain unpunished. Liudolf then conveyed the martyrs' message to the *confratres*, concluding: 'Alas that I ever permitted this!'[68] He then learned that the corpse was being brought to the church, but did not let the *fratres* receive it until a ritual of reconciliation and reincorporation had been performed, as described in Book IV, chapter 71.:

[Liudolf] addressed the dead man angrily, saying: 'How could you be so impudent as to treat carelessly those who are held in great honor by the only begotten son of the living God? And, how have you dared to come here, after such a deed, in the absence of any humble intercession?' The deacon defended the dead man as best he could but received this response from the abbot: 'My beloved brother, you know what this servant has done visibly in your presence, but you are ignorant of what he has done in your absence. I can judge best, knowing that he is presently in grave torment. And now, I will humbly invoke the intercession of our patrons, that through them I may be informed when God has freed this sinner, so that I may grant him absolution and communion with the Church. It is very serious to resist the good; and it is unseemly for mortals to grant indulgence if God is angry.' After these words, the pious abbot came, with bare feet, to that oratory which was his special refuge in difficult situations. He placated God and released the sinner. Then, amid many expressions of gratitude, he arose, remitted the dead man's sin through the power of heaven and before all the brethren, and granted his body communion with the church and burial.[69]

Comparing this narrative to that describing the reform of 1015, as Thietmar at both IV.70 and VII.13 explicitly prompts the reader to do, a stark contrast emerges. In Liudolf's day, the carnal actions of a lone monk led to his exclusion from the spiritual confraternity; but thanks to the intercessions of the pious abbot, he is redeemed from tormented and thereby (spiritually) restored to communion with the *fratres*. Furthermore, Liudolf's actions prioritize a forgiving, inclusive parental love (perhaps even more maternal than paternal) over the strict enforcement of discipline demanded by the more militant martyr-saints. In Thietmar's day, however, Corvey's spiritual and corporeal unity is permanently broken by the strict discipline of a very different kind of leader, who is described neither as abbot nor as pastor, let alone pious or good, but simply by name as Druhtmar.

In the case of Memleben, its *conversio* took the form of its being placed by the emperor under the rule of the abbot of Hersfeld in 1015, thereby losing its previous independence. Thietmar describes these events as follows:

It is with great sadness that we must also note and describe how the monastery at Memleben was deprived of its liberty, so often confirmed in the past, and reduced to servitude. For after Abbot Reinhold had been deposed, and the brothers scattered far and wide, the monastery itself was made subject to the church of Hersfeld and its abbot, Arnold.[70]

Again, the consequences of Henrician reform intervention are, by the *Chronicon*'s account, a disruption of a long-established identity (in this case associated with the monastery's *libertas*) and the dispersal of the confraternity. Thietmar's most focused explication of his opposition to the monarch's strict enforcement of the monastic *regula*, however, occurs in his description in Book VI, chapter 21 of the 1005 reform of

Berge (the monastery where he had been educated) by the archbishop of Magdeburg. There he says of the reformers:

> In fact, those whose new fashion in clothing and manner of life is held to be so admirable are not always what they appear. For scripture says: feigned justice is not really justice at all, but rather double injustice. Of all the fruit of humankind, the most pleasing to God is a good heart which, however, the just may sometimes conceal beneath beautiful, golden clothing or moderation in food and drink. Indeed, who really benefits from that which is taken from abstemious wearers of cloaks? If it contributes to the increase of their churches, the profit is doubled; if undertaken for the sake of God, the souls of the brothers are benefited, but their domestic welfare also benefits from the alms that are attracted. But how can this remain secure if all is displayed outwardly; and if their gain means that many suffer privation? Certainly their exaltation will not last, but rather go sadly to ruin at the end.[71]

According to Thietmar then, contemporary reformers substituted for the established, fruitful emphasis on charitable love and spiritual gifts a new, barren emphasis on the carnal externals of rigid discipline in food, clothing and bodily comportment. The result – as also seen at Fulda, Corvey and Memleben – has been to deprive monks of their monastic *confraternitas* and to ruin the integrity of previously harmonious cloisters.

Thietmar's critical perspective on the Henrician reform is shared by the likewise contemporaneous *Annales Quedlinburgenses*.[72] Reflecting the perspective of the imperial canonesses at Quedlinburg, the *Annales* narrate three examples of Henry II's monastic interventions. The entry for Hersfeld at AD 1004 provides the template that the descriptions of Fulda (at AD 1013) and Corvey (at AD 1014 and 1015) will follow:

> At Hersfeld the monastery suffered at the hands of the king the great deprivation and loss of the ancient law of its fathers, was despoiled of its possessions, was bereaved of its children, and lost in these times – as the wages of sin – those whom it had congregated and educated from the reigning of Charles [Martel] the son of Pippin for 179 years.[73]

Again, the negative impact of reform is expressed in disruption, dispersion and despoliation. The ancient way of the fathers is broken, the collective unity and wisdom of the community are shattered, and the convent loses its *fratres* like a mother robbed of her children. The radical newness of *conversio* is likewise contrasted with the preceding centuries of stability in intellectual, spiritual and material prosperity the community had enjoyed since its foundation in AD 736 (actually 269 years, not 179).[74]

In the case of Fulda, the *Annales* acknowledge that Henry wished to improve the regular discipline of the monks (including their vow of poverty), but that his intervention brought the ancient community not a modest and perhaps welcome correction, but rather a near-total destruction and dispersion. The *cenobites*, rather than being refreshed, have become *vagantes*, the wanderers condemned so strongly as *gyrovages* in the opening chapter of Benedict's Rule.

> The wise King Henry, corrupted perhaps by the council of the foolish, wretchedly plundered the possessions of the monastery of Fulda, because the way of life of the brothers dissatisfied him. The force of the destruction undid the opportunity provided for correcting. Those who had been cenobites, bearing the yoke of Christ, now scattered in every direction as *vagantes*. For 270 years since the first Charles[75] they had served God at that place; now in our times – oh what sorrow! – they were made a spectacle to the world, and to themselves in their suffering, and to the others remaining behind in fear.[76]

Again the tabulation of years back to the monastery's foundation in AD 744 and the rhetoric of the entry itself function together to highlight the disruption of the collective, harmonious monastic life by the emperor's intervention to impose a more militant understanding of the Benedictine Rule upon the community.

The description in the *Annales* of Corvey's reform is divided between two entries, the first of which (AD 1014) describes how Henry II's visit to the cloister turned into a counterproductive reform intervention, which indeed militarized the monks, but materially rather than spiritually:

> In this year the emperor came to Corvey to visit the *fratres*, whose manner of life displeased him, and he wished to correct that monastery by imperial authority. Whence a majority of the monks, defending the institutes of their fathers and raging excessively against the law of the empire, alas, acted foolishly, in a lamentable manner; when they were struck on the cheek, they did not offer the other one like monks, but instead, showing a complete lack of considered

reflection, became rebels and wickedly offered a fight. What then took place, and in our times, was more stupefying than the pen can record; but seventeen of them, being captured were taken away to be imprisoned, while the others followed the commands of the emperor.[77]

Where the monks of Stavelot, as described by Everhelm, had allowed their lay co-conspirators to take up arms against Poppo, the Quedlinburg *Annales* portrayed the monks of Corvey as taking up weapons themselves. Caught between the *instituta patrum* and the *ius imperii*, the *fratres* could have chosen the path of humility, turning the other cheek following the Gospel commandment; but this would presumably have resulted, as at Hersfeld and Fulda, in the devastation of their community. So they chose the even worse option of trying to defend their monastic *habitus* by force, through violence violating those very *instituta* and thereby destroying what they had hoped to preserve. In the end, the community still was broken, with monks now expelled not as unjustly dispersed wanders, but as justly imprisoned criminals.

Because of this, when at the following year (AD 1015) the annalist narrates how Henry II finally deposed Abbot Wal and installed Druhtmar of Lorsch in his stead, the monks of Corvey are no longer simple victims, and the reformer is not necessarily a villain. Still, an ancient tradition has been terminated, and what has replaced it is of questionable value:

The emperor then once again paid a visit to Corvey, and forcefully changed the privileges and many ancient customs which the monks had held for 239 years since they had been granted by Louis the Pious; also removing the father of the monastery, the emperor introduced one unknown and perhaps good, who would correct errors as though he possessed better knowledge, and would teach those who had deviated from the path of the holy rule to march forward more warily. Afterward there were those who greatly bemoaned and lamented the unworthiness of his way of life, a way of life which to almost all the other monks shone forth as a clear example. The dissatisfied *fratres* exhorted one another that it was better to disperse than to submit themselves to this outrage. And thus it happened, that a few persevered in this purpose, some occupying themselves with wretchedly wandering through the world, but most returning again by the grace of God, since it seemed better to submit themselves to the rule than to prize things without worth.[78]

In the end, the Quedlinburg *Annales* seems to conclude, obeying an abbot who might or might not be learned, and might or might not be good, is preferable to wandering lost in the world outside the monastic discipline, but the privileges and *consuetudines* which had held at Corvey for 239 years (actually 193 years, since the *Annales* again add incorrectly) were better. When an abbot who can be called father is replaced by 'someone unknown' who demands (rather than elicits) obedience, the loving community of the *familia* (where authority is both maternal and paternal) has been lost to loveless order of an exclusively masculine military discipline of command and obedience which drives large numbers of monks away.

Finally, Hermann of Reichenau's chronicle (written in the 1040s and early 1050s) comes from a *Reichskloster* which had experienced the Henrician reform in 1006.[79] One of the main themes of the work is the unbroken continuity of the monastery's traditions and privileges beginning with its foundation in 724, when Saint Pirmin received the island of Reichenau from Charles Martel, drove out the serpents, and instituted the cenobitical life. Immo of Gorze's imperially imposed abbacy of 1006–1008 is the stark exception to this theme of continuity. When narrating Henry II's reform intervention at Reichenau, Hermann portrays Immo as an interloper whose strict discipline had a destructive impact on the community:

> AD 1006. With death of Abbot Werinhar of Reichenau, the brothers elected the monk Henry. But King Henry II, detesting his insolence although he had accepted money from him, and also made hostile through accusations brought to him against the *fratres*, instead put in place – despite the unwillingness of the *fratres* themselves – a certain Immo, the abbot of Gorze and a harsh man who also at the same time held the monastery of Prüm. For this reason, not a few of those same *fratres* voluntarily left that place, while others indeed were gravely afflicted by that man through fasts, beatings, and exile, and to pay the penalty for its sins the noble monastery suffered a grave loss in highly valued men, books, and treasures of the church: just as Rupert, a monk both noble and brilliant in learning, the paternal uncle of my mother, tearfully lamented in prose, rhythm, and meter.[80]

Hermann's criticism of the reform begins with a censuring of Henry II hardly veiled by his seeming to acknowledge possible faults among the monks. Accusations were made against the *fratres*, but in the context both of the remainder of this entry and of the *Chronicon* as a whole, he presents these as certainly without warrant. The abbot-elect had given money to the king, but again Hermann's description suggests that he is actually saying Henry II had accepted the customary gift in exchange for recognizing Henry of Reichenau's election, only to then renege and in effect turn that gift into straightforward *pecunia*. The *fratres* reluctance to accept Immo is shown to be perfectly justified, while the latter's willingness to hold three abbeys at once (whereas the monks had rightly preferred to elevate a pastor from among their own ranks) calls his commitment to *confraternitas* seriously into question. The consequences of Immo's reform bear this out. The harshness of the new discipline, including beatings and fasts, leads to the fracturing of *communitas*, some *fratres* leaving immediately and others being driven away into exile by the excessive physical rigors imposed under the new regimen. With loss of community solidarity comes a loss of learning and of treasures both spiritual and material, exemplified by the monk Rupert whose nobility, brilliance and literary skill represent both figuratively and – in his (unfortunately lost) poetic lament – literally the extent of the bereavement threatened to Reichenau by Immo's abbacy.

By contrast, Hermann represents the arrival of Bern in 1008 as the monastery's salvation, the restorer of Reichenau's traditions and *confraternitas*:

> AD 1008. In this same year King Henry, the cruelty of Immo at last being known after two years, removed that man and placed Bern – a pious and learned man and a monk of Prüm – as abbot of Reichenau; who, joyously received, collected the dispersed *fratres* together again and was consecrated twenty-ninth abbot of the same place by Bishop Lantpert of Constance; distinguished by great learning and piety, he ruled for forty years.[81]

With his spiritual and intellectual gifts, Bern's leadership heals the rupture instigated by Immo – on whom, pointedly, Hermann *never* bestows the title 'abbot of Reichenau', whereas Bern's accession to the title is here mentioned twice. The abbatial succession is renewed, the congregation is reassembled and the cloister's traditions of scholarly and spiritual distinction are reinstated, while even the length of Bern's time in office is used to emphasize the restoration of a stable, durable order. Instead of ruling by paternal might and discipline, as Immo had attempted to do, Bern's authority (parental, or even maternal, rather than militant or restrictively masculine) depends upon the love and joy he evokes in the dispersed *fratres*, who willingly return to the fold under his devoted care.

Like Thietmar and the Quedlinburg annalist, Hermann represented the Henrician reform as abrupt and radical, harsh and militant, and shattering and senseless, and its obliterative impact is contrasted starkly with the collective spiritual and intellectual heights which generations of *fratres* had achieved through their unified adherence to the traditional customs and habits of their *coenobium*. At the heart of the ideological struggle between the supporters and opponents of Henry II's monastic reform agenda was a debate about the appropriateness of a military metaphor for understanding monastic service and abbatial leadership. Henry II called for monks and nuns to battle more effectively for God under the order of the Benedictine Rule, to become more virile both in the quality of their leadership and the discipline of their service. Although, significantly, the chief architects of this reform were all men, they called upon *sanctimoniales* to take up arms on the same terms as *fratres*. The goal of this particular initiative was not to exclude women from monastic service, or to differentiate their service from that of men, but to summon both monks and nuns to a more masculine militancy. The highest command remained in the hands of biological males, but these men welcomed their fellow abbesses into the officer corps. Similarly, the canonesses of Quedlinburg were as involved in resisting this agenda, and in the same terms, as Hermann of Reichenau or Thietmar of Merseburg.

The two principle issues in the debate were leadership and *confraternitas*. In the conceptual framework of the Henrician innovators, these qualities needed to be more intensively masculinized. Specific abbots, heroic and disciplined individuals, were to assume direction of this project – where necessary, by assuming management of multiple abbeys simultaneously. The success of these 'great men' was then commemorated in *vitae*, using a variant of the hagiographic genre as a medium for communicating the unique talents and vision of a Poppo of Stavelot or Godehard of Niederaltaich. The congregation of monks, in turn, was to serve obediently and humbly under these commanders; if the congregation resisted submission, it would be cowed into submission and punished – whether by God, the abbot or both. For the opponents of Henry II's reform interventions, monastic leadership and *confraternitas* needed instead to be more strongly collectivized. Not the abbot, but the community as a whole represented the intellectual, artistic, spiritual attainments of a monastery united together in love (*caritas*). That community, in turn, encompasses not only the monks of the present, but all the generations who have lived in the cloister since its foundations and have followed the shared *habitus* as defined by the original members, the *fratres* of earlier days now serving as the collective *patres* of the brothers in the present. This vision, at once communal and nostalgic, found *its* literary expression in the form of the chronicle, which valued (and evaluated) the present as an extended continuation of the past. In these sources, the would-be reform abbots are presented as headstrong and harsh men who presumptuously and self-deludedly claimed to have better wisdom than the ancient, communal knowledge of the monasteries they were trying to 'renew'. As a result, their individualistic pride and bellicosity destroyed the treasures of confraternity and collaboration.

The Henrician reform ideology was explicitly masculine in its embrace of the warrior model of soldiers and commanders; the position of the nostalgists defined itself against that position by avoiding exclusively masculinist descriptions of claustral life. Though the terms *patres* and *fratres* featured in their discourses, for the most part these simply reflected the biological sex of the monastics being discussed, and were combined with strongly maternal imagery in descriptions of ideal abbots like Liudolf. While the ideological and social implications of every use gendered language should be critically examined, in this instance it should also be recognized that the ideological position adopted by Thietmar, the Quedlinburg *Annales* and Hermann did *not* rely primarily upon a strategic use of masculine or feminine discourses. Instead, it deployed a *collectivizing* discourse whose impact relied on the appeal of a community of harmonized variety (shared intellects, shared love and shared traditions) rather than on the appeal of being disciplined to a single virile norm. In the monasteries of early eleventh-century Germany, the dispute about claustral reform did not pit two interpretations of clerical masculinity against one another, but instead juxtaposed Henry II's order to be more manly with a call to resist and remain true to the ancient, harmonious *communitas*.

Notes

1. Joachim Wollasch, 'Monasticism: The First Wave of Reform', in *The New Cambridge Medieval History*, vol. 3: *c. 900–c.1024*, ed. Timothy Reuter (Cambridge, 1999), 163–85; see 166–7 and 174–80 on Cluny's uniqueness in being 'free of all spiritual and temporal *dominatio*'.
2. Wollasch, 169–74 (on tenth-century examples of these partnerships) and 180–5 (on the era of Henry II).
3. The key works on Henry II's monastic reform initiative are Hubertus Seibert, 'Herrscher und Mönchtum im spätottonischen Reich. Vorstellung – Funktion – Interaktion', in *Otto III.–Heinrich II.: Eine Wende?*, ed. Bernd

Schneidmüller and Stefan Weinfurter (Sigmaringen:1997), 205–66 and Hartmut Hoffmann, *Mönchskönig und rex idiota: Studien zur Kirchenpolitik Heinrichs
II. und Konrads II.* (Hanover, 1993), 27–49. Both Seibert and Hoffmann agree
on the distinctiveness of Henry II's cenobitical reform project, and distinguish his vision of monastic reform from those of his predecessor Otto III
and successor Conrad II.

4. Vern L. Bullough, 'On Being Male in the Middle Ages', in *Medieval Masculinities: Regarding Men in the Middle Ages*, ed. Clare A. Lees (Minneapolis, 1994),
   31–45, quoting David D. Gilmore's definition from *Manhood in the Making:
   Cultural Concepts of Masculinity* (New Haven, Conn., 1990), 34.

5. Jacqueline Murray, 'Masculinizing Religious Life: Sexual Prowess, the Battle
   for Chastity, and Monastic Identity', in *Holiness and Masculinity in the Middle
   Ages*, ed. P.H. Cullum and Katherine J. Lewis (Cardiff, 2004), 24–42, quoted
   at 25.

6. Ibid., 33.

7. Dyan Elliott, *Fallen Bodies: Pollution, Sexuality, and Demonology in the Middle
   Ages* (Philadelphia, 1998) provides a fascinating overview of the sexualiza-
   tion of the clergy as a result of the obsession with priestly celibacy in the
   discourses of later medieval ecclesiastical reform.

8. See for instance Jo Ann McNamara, 'The *Herrenfrage*: The Restructuring of the
   Gender System, 1050–1150', in *Medieval Masculinities*, 3–29, which argues
   that the subjection of the secular clergy to celibacy 'aimed at a church virtually free of women at every level but the lowest stratum of the married
   laity' (quoted at page 7); or, on the other hand, R.N. Swanson, 'Angels Incarnate: Clergy and Masculinity from Gregorian Reform to Reformation', in
   Masculinity in Medieval Europe, ed. Dawn M. Hadley (London and New York,
   1999), 160–77, which argues that celibacy of the late medieval clergy placed
   them in a condition of gender liminality which led them to make common
   cause with women 'to civilize and Christianize [sexually active] medieval
   lay men . . . [in] a conspiracy [of women and emasculine clerics] against male
   control over the family and domestic life' (quoted at page 170).

9. Maureen Miller, 'Masculinity, Reform, and Clerical Culture: Narratives of
   Episcopal Holiness in the Gregorian Era', *Church History* 72 (2003): 25–52,
   quoted at page 28.

10. Ibid., e.g. 34 and 41–2.

11. Ibid., 46.

12. Ibid., 45 and 44; also see 38–9.

13. Ibid., 37.

14. Ibid., 28, 41 and 49.

15. J.L. Nelson, 'Monks, Secular Men, and Masculinity, c. 900', in *Masculinity
    in Medieval Europe*, 121–42, argues that laymen in positions of authority
    were measured (and measured themselves) not only against the model of
    the warlike David but also against the contrasting example of his son, the
    philosopher-ruler Solomon; Emma Pettit, 'Holiness and Masculinity in Ald-
    helm's *Opus Geminatum De virginitate*', in *Holiness and Masculinity*, 8–23,
    argues that the seventh-century Anglo-Saxon monk and hagiographer dif-
    ferentiated between physical potency and the interior battle against vice,
    defining the latter as the more spiritual form of masculinity in which both

male and female saints participated; Dawn Marie Hayes, 'Christian Sanctuary and Repository of France's Political Culture: The Construction of Holiness and Masculinity at the Royal Abbey of Saint-Denis, 987–1328', in *Holiness and Masculinity*, 127–42, argues that the Capetian kings and Saint-Denis monks participated together in a 'manly' partnership in which questions of sexuality or celibacy were subordinate to the shared goal of defending the peace, prosperity and stability of the French *familia* from all threatening and disruptive forces.

16. Hoffmann, *Mönchskönig*, 27; see also Seibert, 'Herrscher', 215, 234, 238, etc.

17. See John W. Bernhardt, 'Der Herrscher im Spiegel der Urkunden: Otto III und Heinrich II. im Vergleich', in *Otto III. – Heinrich II.: Eine Wende?*, ed. Bernd Schneidmüller and Stefan Weinfurter (Sigmaringen, 1997), 327–48. and Hartmut Hoffmann, 'Eigendiktat in den Urkunden Ottos III. und Heinrichs II', *Deutsches Archiv* 44 (1988): 390–423.

18. MGH *Diplomata regum et imperatorum Germaniae: Heinrici II. et Arduini diplomata* [hereafter DD HII], no. 277.

19. For Henry II's general policy of political centralization in the empire, see e.g. Stefan Weinfurter, 'Die Zentralisierung der Herrschaftsgewalt im Reich unter Heinrich II', *Historisches Jahrbuch* 106 (1986): 241–97.

20. Seibert, 'Herrscher und Mönchtum', 236 note 159.

21. DDH II, no. 370 (July 1017).

22. For example, MGH *Diplomata regum et imperatorum Germaniae: Conradi I. Heinrici I. et Otto I. diplomata* [hereafter DD CI, DD HI and DD OI], DD CI, no. 6 (for Fulda); DD HI, nos. 1 (for Fulda) and 3 (for Corvey); DD O1, nos. 2 (for Fulda), 121 (for Weissenberg), 167 (for Stavelot) and 233 (for Ellwangen). Also MGH *Diplomata regum et imperatorum Germaniae: Otto II. et Otto III. diplomata* [hereafter DD OII and DD OIII], DD OII, no. 103 (for Fulda) and 219 (for Stavelot).

23. For Einsiedeln: DD OI, nos. 94, 218 and 275 and DD OII, no. 123; for Saint-Remi: DD OI, no. 286 and DD OIII, no. 122.

24. DD OIII, no. 152 (Petershausen); nos. 157, 158 and 161 (Waldkirch); no. 262 (Luisberg); no. 318 (Seeon).

25. DD OIII, no. 152, at lines 25–6.

26. DD HII, no. 18 (28 September 1002): 'monachos iuxta regulam beati Benedicti militantes.'

27. DD HII, no. 58 (22 October 1003): 'institutis ibi sacris virginibus sub norma sancti patris Benedicit deo militantibus viriliter et humiliter servientibus.'

28. DD HII, no. 77 (17 June 1004): 'servisque dei secum per succedencia tempora in prefata cella deo ad regulam sancti Benedicti militantibus.'

29. DD HII, no. 83 (8 August 1004): 'ad augmentum loci et fratrum ibidem deo sub regula sancti Benedicti militantium.'

30. DD HII, no. 230 (26 June 1011): 'idem monasterium ex integro in potestatem abbatissę et sanctimonialium ibidem deo sub regula sancti Benedicti commilitantium revertatur.'

31. DD HII, no. 375 (for Kaufungen: 6 December 1017): 'in quo [loco] virgines sub regula sancti Benedicti ordinavit'; no. 413 (for Fulda: 1 July 1019): 'ad utilitatem tantummodo fratrum et aecclesiae ibidem sub regula sancti Benedicti deo servientium.'

32. DD HII, nos. 420 (for Kaufungen: 31 December 1019), 428 (for Göss: 1 May 1020) and 446 (for Dietkirchen: 10 August 1021); also nos. 421 (for Abdinghof: 18 February 1020), 431 (for Tegernsee: 29 May 1020), 434 (for Prüm: 27 September 1020), 441 (for St. Emmeram: 3 July 1021) and 503 (for Heiligenberg: 13 December 1023).

33. Nelson, 'Monks, Secular Men, and Masculinity'. For a contemporary description of Henry II as both David and Solomon, see Bern of Reichenau's letter no. 4 in *Die Briefe des Abtes Bern von Reichenau*, ed. Franz-Josef Schmale (Stuttgart, 1961), 22–4.

34. See note 33 above.

35. Wolfer, *Vita Godehardi episcopi prior*, MGH SS 11, 167–196 [hereafter *Vita Godehardi*]. Twenty years later, Wolfher wrote a second biography of Godehard which will not be considered here.

36. Ibid., 168 (prologue).

37. Ibid., c. 2, 171.

38. Ibid., c. 13, 177.

39. Ibid.

40. Ibid., c. 14, 178.

41. Ibid.

42. Everhelm, *Vita Popponis abbatis stabulensis*, MGH SS 11, 293–316. Actually, while the content of the *vita* came from Everhelm, he appears to have assigned most of the task of composition to his own subordinate, the Hautmont monk Onulph: see e.g. Philippe George, 'Un moine est mort: sa vie commence. *Anno 1048 obiit Poppo abbas Stabulensis*', *Le Moyen Âge* 108 (2002): 497–506, at pages 497–8 and 506.

43. Everhelm, *Vita Popponis*, c. 1, 294–5.

44. Ibid., c. 2, 295.

45. Ibid.

46. Ibid., 295.

47. Ibid., c. 9, 298–9.

48. Ibid., c. 13, 301.

49. Ibid.

50. Ibid.

51. Ibid., c. 15, 302.

52. Ibid.

53. Ibid., c. 16: 'ubi [i.e. St. Maximin] in eum similis, immo gravior quorundam cum insidiandi copiis invidia processit; eorum videlicet, quos ex monasticis regularibusque observationibus via per abruptum deflexit.'

54. Hoffmann, *Mönchskönig*, 46–9; Seibert, 'Herrscher und Mönchtum', 220–2 and 224; Joachim Wollasch, 'Kaiser Heinrich II. in Cluny', *Frühmittelalterliche Studien* 3 (1969): 327–42; idem, 'Cluny in Deutschland', *Studien und Mitteilungen zur Geschichte des Benediktinerordens und seiner Zweige* 103 (1992): 7–32.

55. DD HII, nos. 69 (for Peterlingen), 251 (for S. Pietro in Ciel d'oro in Pavia: in 1012), 288 (for S. Maria at Farneta: in 1014) and 399 (for Leno: in 1019).

56. Hoffmann, *Mönchskönig*, 46–7 with references.

57. Iotsald of Saint-Claude, *Vita des Abtes Odilo von Cluny*, ed. Johannes Staub, MGH SrG 68 (Hanover, 1999), I.8, 162; Rodulphus Glaber, *Historiarum*

Libri Quinque: The Five Books of Histories, ed. John France (Oxford, 1989), I.5.23, 40.

58. *Liber tramitis aevi Odilonis abbatis*, ed. Peter Dinter as volume 10 of *Corpus Consuetudinum Monasticarum* (hereafter CCM), ed. Kassius Hallinger et al. (Siegburg, 1980), 285.

59. See Hoffmann, *Mönchskönig*, 47–8.

60. MGH *Poetae* 5, No. 38/I, 394–5.

61. Thietmar of Merseburg, *Chronicon*, ed. Robert Holtzmann, MGH SrG, n.s. 9 (Berlin, 1935); translated into English by David A. Warner as *Ottonian Germany: The Chronicon of Thietmar of Merseburg* (Machester and New York, 2001).

62. Warner, 298; Holtzmann, 383, 385.

63. Warner, 123; Holtzmann, 92.

64. Warner, 316; Holtzmann, 412.

65. Thietmar, II.18 and III.4: Warner, 105 and 129; Holtzmann, 58, 60 and 100.

66. Thietmar, III.9: Warner, 133–4; Holtzmann, 106, 108.

67. Warner, 134; Holtzmann, 108.

68. Warner, 200; Holtzmann, 210, 212.

69. Warner, 200–01; Holtzmann, 212, 214.

70. Warner, 329; Holtzmann, 436.

71. Warner, 251; Holtzmann 298, 300.

72. *Die Annales Quedlinburgenses*, ed. Martina Giese, MGH SrG 72 (Hanover, 2004).

73. Ibid., 522.

74. Ibid., 421.

75. The annalist means Charles Martel, though the foundation actually dates to the time of Pippin the Short.

76. Ibid., 539–40.

77. Ibid., 543.

78. Ibid., 546.

79. Hermann of Reichenau, *Chronicon*, MGH SS 5: 74–133.

80. Hermann of Reichenau, *Chronicon*, a. 1006, in MGH SS 5, 118.

81. Ibid., 119.

4
Spiritual Warriors in Citadels of Faith: Martial Rhetoric and Monastic Masculinity in the Long Twelfth Century

Katherine Allen Smith

When Robert of Châtillon fled the rigors of the Cistercian Order for the comparative comforts of Cluny, his kinsman and former mentor Bernard of Clairvaux (d. 1153) composed an open letter of reproach intended to demonstrate the superiority of the Cistercian way of life to the errant monk as well as the leaders of his new Order.[1] Writing to Robert at his new home, Bernard spoke not merely as a monastic superior to an apostate but as a military commander to a soldier, and as one man to another. Bernard insists that in leaving Clairvaux, Robert has deserted his army of fellow monks in the midst of a great battle, exchanging 'the arms of fighting men' for 'comforts for the weak', namely soft clothing and abundant food, and leaving himself open to accusations of cowardice and weakness. It is Robert's sworn duty, Bernard reminds him, to return and fight beside his brothers in arms, the monks of Clairvaux:

> Arise, soldier of Christ (*miles Christi*), arise! Shake off the dust and return to the battle from which you fled ... Do you think that because you fled the line of battle you will escape the enemies' clutches? The enemy pursues you more eagerly when you flee than he would one who fought back, and more daringly ambushes you from behind than he would oppose you to your face. Are you safe, now you have thrown down your arms, and are sleeping away the morning, even at that hour when Christ rose again? Do you not know that without arms you are more timid and less intimidating to your enemies? A host of soldiers is besieging the house, and you are sleeping? They are already climbing the walls, destroying the barriers, charging through the rear gates. Would you not be safer with others than alone? Would

you not be safer armed in camp than lying naked in bed? Go, take up arms, and flee to your fellow soldiers (*commilitones*) whom you deserted by running away.[2]

Bernard used this vivid martial rhetoric to appeal to his cousin as both a monk and a man,[3] and suggest a close relationship between warfare and the monastic ideal. As constant readers of the Scriptures, the Fathers and Benedict's *Rule*, monks like Bernard were steeped in the language of spiritual warfare, and viewed themselves as *milites Christi*, or 'warriors of Christ' who traced their lineage back to the earliest monks and beyond, through the martyrs of the early Church to Christ himself. Further, in common with many monks of their day, Bernard and Robert were born into families whose status and wealth depended directly on war, and would have spent their earliest years absorbing tales of the battlefield. Monastic men's ideas about war were thus shaped by two main sources: sacred texts, especially Scripture, works of hagiography and the liturgy, which treated warfare in historical and allegorical terms; and the social world of medieval Christendom, in which warfare was the calling of violent men whose relationship with the Church was fraught with tensions. In the time period under consideration here, most men who entered monastic communities came from the ranks of families of *bellatores*, the martial elite then coalescing into a self-consciously defined noble class.[4] For men from such backgrounds, especially the growing number whose conversions to the religious life occurred in adulthood,[5] speaking the language of war may have allowed them to identify with masculine ideals familiar to them from their earlier lives. But martial rhetoric also appealed strongly to monastic authors, such as Peter Damian, who came from humble backgrounds, as well as those who, like Orderic Vitalis, spent their lives from childhood onward in the cloister.[6] How, then, are we to understand why monastic men so often turned to war as a source of metaphors and moral lessons in their letters, sermons and hagiography, and what might this proclivity suggest about monastic masculinity?

Ruth Mazo Karras has argued that clerical men who employed the sort of military metaphors discussed below 'were not so much trying to live up to a secular model [of masculinity] as they were transcending it'. Rather than competing with a knightly masculinity based on aggression and martial skill, clerics used familiar masculine attributes to formulate a distinct concept of masculinity, one which celebrated an ideal of a powerful, yet celibate, male who successfully 'fought' and 'conquered' his flesh with God's help.[7] Such a model incorporated elements, like an emphasis on spiritual struggle, that had characterized clerical masculinity since the early Church, even while it drew upon and challenged contemporary secular notions of gender. My own reading of monastic masculinity in the long twelfth century views the masculinity of monks as deeply rooted in the activities, values and aesthetics that defined the monastic *habitus*. At the same time, the gender identity of monks hardly developed within a cloistered vacuum, as their self-perception was influenced by external agents, such as the ideas of church reformers, and the values inculcated in them by relatives and peers during their years in the world. While the masculinity of twelfth-century monks was by no means passively shaped in the image of a hegemonic knightly or even reformed clerical masculinity, it engaged in dialogue with both of these models.

Mindful of the historical associations of spiritual combat, monastic men used martial rhetoric to insert themselves into a genealogy stretching back to the most ancient Christian traditions and beyond, which included the ancient Israelites, martyrs and Desert Fathers whom they perceived as the first great warriors of the Church. In its privileging of heroic male 'ancestors', such a spiritual lineage might be compared to the agnatic genealogies commissioned by elite families beginning in the eleventh century.[8] But martial rhetoric also helped monks navigate the world of the present, and define their place within social, spiritual and gender hierarchies. At a historical moment when the Church's power increasingly depended on its ability to harness the violence of lay warriors,[9] and monks' ancient and traditionally exclusive self-identification as soldiers of Christ (*milites Christi*) was being challenged from several quarters,[10] monastic writers turned to *rhetorical* violence to claim a different type of authority for themselves.[11] The authority claimed by monastic writers who engaged in rhetorical combat was of a spiritual nature, and asserted itself to be superior to any mere earthly power possessed by lay warriors, even those who fought in the service of the Church, but it was also a masculine authority, by virtue of the strongly masculine connotations of military endeavor and martial virtue in medieval Europe. Thus, while monks cannot be said to have been simply co-opting an aristocratic warrior masculinity, since their identity as *milites Christi* predated the existence of the knightly class, through its emphasis on martial virtues monastic conceptions of masculinity engaged with ideas of manliness (and unmanliness) central to the gender identity of contemporary lay warriors.[12] But monks' insistent presentation of themselves as virile spiritual warriors did not, I would argue, reflect doubts about their status as 'real men' in comparison with lay warriors so much as a confidence in their own prowess in a form of combat far more dangerous and demanding than any fought on earthly soil.

Though images of knightly masculinity and realistic descriptions of secular combat fill the texts considered below, the vivid martial rhetoric employed by monastic writers of this period was a specialized discourse used most often by monks when writing to, for, or about men like themselves. Modern scholars have cautioned against reading symbolism in monastic writing as evidence of lived experience, and the martial rhetoric found in the texts considered here undoubtedly reflects not so much monastic authors' actual experiences with warfare (though it may reflect these as well) as it does their self-perception as monks and men.[13] Verbal images of spiritual combat conjured up in letters, sermons and hagiography reinforced a model of the ideal monk as a strong, courageous, unambiguously masculine warrior, and created for monastic men a sense of membership in a cohort of virile brothers-in-arms. A careful reading of martial rhetoric reveals monks self-consciously writing and acting as men, and implicitly gendering male many activities central to monastic life. While religious women might be labeled *milites Christi*, and even be well versed in the uses of martial rhetoric themselves, monks persistently associated the model spiritual warrior – and by extension the model monastic individual – with qualities and activities strongly coded as masculine within medieval culture.

I Spiritual warfare in the Christian tradition

Spiritual warfare has long been recognized, and the early genealogy of this theme fully explored, even while the abundant martial rhetoric to be found in later monastic texts has received relatively little attention.[14] The earliest Christians inherited the concept from the Hebrew Scriptures as well as the Gospels, but the topos of the *miles Christi* was more fully developed by Paul of Tarsus, whom the eleventh-century ascetic Peter Damian affectionately nicknamed 'the spiritual drillmaster'.[15] The martyr movement took up Paul's call to arms, even as church leaders questioned whether Christians ought to serve in the Roman army. In his *Ad martyras*, Tertullian (d. 230) urged those awaiting execution to see their prison as a training ground, reminding them that they 'have now been called to the army of the living God', and must accordingly be prepared to exchange warmth for cold, soft clothes for armor and quiet for tumult as long as Christ chose for their battle to last.[16] The *miles Christi* of the early Church was an unambiguously masculine ideal, so much so that Late Antique writers avoided using the title in reference to female subjects; although female martyrs might do spiritual combat in the arena, in doing so they displayed a distinctively manly *virtus*.[17] In the words of Mathew Kuefler, the concept of spiritual warfare 'was turned into the heart of a sophisticated (and complicated) defense of Christian manliness' against Roman pagans who mocked as unmanly Christians' pacifistic acceptance of suffering. According to the ideology of spiritual combat elaborated by early theologians and hagiographers, the holy men who laid down their arms were far stronger and braver than pagan persecutors who gave in to unmanly rage or Roman soldiers adhering to the traditional *vita militaris*.[18]

Professing themselves the direct heirs to the martyrs, the progenitors of the Christian monastic tradition adopted the language of spiritual combat to describe their travails in the desert.[19] Glossing Paul (2 Tim 2:3), the third-century theologian Origen advanced the idea that a select few Christians might fight on behalf of the whole body of the faithful, wielding the weapons of prayer, fasting and chastity, and monastic writers increasingly identified themselves with this military elite.[20] The writings of the ascetic priest Jerome (d. 420), which became staples of later monastic libraries, are filled with martial metaphors. In a letter from which Bernard of Clairvaux would later borrow imagery,[21] Jerome reproved Heliodorus, a former soldier who had come to regret his conversion to monastic life, as a deserter from Christ's army. Jerome warned Heliodorus that as a monk he had no business living as a 'pampered soldier (*delicate miles*)' in his father's house, where worldly comforts would inevitably destroy his martial spirit and effeminize him:

> A body that is used to a tunic cannot support a cuirass, a head that has worn a linen hood shrinks from a helmet, a hand idleness has softened is galled by a hard sword-hilt. Hear your king's proclamation: 'He that is not with me is against me, and he that gathers not with me, scatters (Mt 12:30).'[22]

By Jerome's day, the assumption that the monk was the true *miles Christi* was firmly established, and the association of Christian *virtus* and manliness first made in connection with the martyrs had found its way into discussions of monastic exemplarity. David Brakke has shown how descriptions of spiritual combat in fourth- and fifth-century hagiography portray monastics of both sexes as relatively masculine or feminine depending on their prowess in single combat against demonic adversaries. Just as a holy woman might display a spiritual manliness through displays of ascetic fortitude, it was imperative that monks avoid being effeminized by succumbing to temptation.[23]

The sixth-century *Rule of Benedict*, which would serve as the basis of communal life for most monasteries in the West in the Central Middle Ages, envisioned the monastic community as a militia and the brethren's main occupations as battles.[24] As he declared in his Prologue, Benedict wrote with an audience of monastic soldiers in mind:

> This message of mine is for you, then, if you are ready to give up your own will, once and for all, and armed with the strong and noble weapons of obedience to do battle (*militare*) for the true King, Christ the Lord.[25]

For Benedict, the struggle to overcome the will – the central struggle of monastic life – was a battle each monk must fight within himself. At the same time, he added, 'We must prepare our hearts and our bodies for the battle of holy obedience (*sanctae…oboedientiae militanda*)', a battle which soldier-monks will collectively fight under the generalship of Christ, or Christ's earthly representative, their abbot.[26] The collective warfare Benedict spoke of was fundamentally different from that celebrated by the martyrs, early desert ascetics, or the hermits of his own day, men who (in the words of the *Rule*) had shown themselves ready to 'go from the battle line (*acies*) in the ranks of their brothers to the single combat of the desert'.[27] As spiritual beginners, Benedict's monks were presumed to be novices in the daily battle that would consume their future lives, and would gain experience in war by fighting as part of a cohort of *milites Christi* and acquiring thereby the soldierly virtues of obedience to superiors and loyalty to brothers in arms.[28]

While early medieval authors did not use martial rhetoric exclusively in connection with men, they consistently associated spiritual warfare with qualities like physical strength, courage and fortitude which were commonly defined as manly, even in reference to female subjects. For hagiographers who associated sanctity with bellicose struggle, holy women might make capable spiritual soldiers provided that they assumed the masculine qualities required of victors. In the preface to his *vita* of the Frankish nun Monegund, the sole woman awarded a place in his *Liber Vitae Patrum*, Bishop Gregory of Tours (d. 594/5) explained that God

> gives us as models not only men, but even the lesser sex, who fight not feebly, but manfully (*viriliter*); He brings into His celestial

kingdom not only men, *who fight as they should*, but also women, who exert themselves in the struggle with success.[29]

In other words, like ordinary warfare, spiritual combat was the natural province of men, who 'fight as they should'. Women warriors were exceptions, admirable because of the virile strength that sets them apart from most members of their sex. As we shall see, the gendering of spiritual warfare as a naturally masculine activity remained a feature of later monastic thought.

II Speaking the language of war

A survey of monastic writings of the eleventh and twelfth centuries turns up military metaphors, often quite elaborate ones, with astonishing frequency. Modern readers have often noted this martial rhetoric, but have only begun to connect it to larger issues of monastic and masculine identity.[30] A growing interest in spiritual warfare after 1000 may reflect currents within monasticism, as members of new Orders emphasizing asceticism and influenced by eremitic traditions naturally drew inspiration from the heroic martial traditions of Late Antiquity. But monastic fascination with spiritual combat likely also derived from a growing preoccupation with temporal warfare as a marker of boundaries – between worldly power and spiritual authority, men of the sword and men of God. As reformers sought to restructure the ecclesiastical hierarchy and redefine its relationship to secular leaders, debates swirled over definitions of just war, and church leaders edged closer to definitive rulings on when and against whom war could be waged, and which members of society were permitted to bear arms and take part in battle. One result of these debates was the classification of regular and secular clergy, along with women and the poor, as groups whose inability to defend themselves by force of arms gave them a special claim on the protection of Christian warriors.[31] Out of reformers' heightened concern with clerical purity developed a new clerical masculinity, dubbed by Maureen Miller 'an extreme masculinity', which self-consciously defined clerics as more manly than secular men of all ranks, by virtue of the purity they attained through avoidance of polluting agents such as female bodies, money and arms.[32] Unlike bishops and priests, regular clergy had long been expected to avoid such pollutants, but their regular inclusion in contemporary ecclesiastical legislation aimed at curbing clerical violence indicates that fighting monks were by no means unknown.[33]

Even while clerics were denied actual participation in war by a reform-ing ecclesiastical hierarchy, spiritual battles raged more fiercely than ever within the monasteries of Christendom. Churchmen who con-demned temporal combat in the strongest terms employed the language of spiritual warfare to express admiration of their peers and assure one another that, though they might be classified as non-combatants, they were by no means lacking in virile strength. Equating military service (*militia*) with wrongdoing (*malitia*), Anselm of Canterbury (d. 1109) entreated a young knight to forsake 'the bloody confusion of wars' and to become a soldier of Christ at the abbey of Bec.[34] Anselm often addressed male monastic correspondents (but never laymen or women) as fellow soldiers, enquired after the progress of their spiritual battles and congratulated novices on their recruitment into Christ's army.[35] Tales of the spontaneous conversion of secular warriors into *milites Christi* suggest that the line between the two kinds of soldiers could be a very thin one, at least in the monastic imagination. This is illus-trated by a story in the *Vita Prima Sancti Bernardi*, which describes how Bernard of Clairvaux's prayers brought about the conversion of a group of young knights 'given over to worldly chivalry', who had stopped at his abbey on their way from one tournament to another. Having drunk ale blessed by the holy abbot, however, the men fell under his spell: they rode off with the intention of continuing on to another of 'those detestable gatherings popularly known as tournaments', but soon felt compelled to return to Clairvaux and 'dedicate their sword arms to the pursuit of spiritual warfare'.[36] The conversion of adult warriors to the religious life was often construed in similar terms, in order to empha-size that success in both callings depended on the possession of the very same masculine qualities, and to promote the idea that only the most virile knights were drawn to the better warfare of the cloister.

Proponents of the great eleventh-century eremitic revival enthusias-tically took up the ancient theme of the solitary as *miles Christi*, and hermits and members of new semi-eremitic Orders became renowned for their martial exploits. The men who fled to the remote forests and mountains where these communities sprang up strongly identified with the martial heroics of the martyrs and desert ascetics, and strove to equal their spiritual forefathers in courage and virile resistance to temp-tation.[37] The letters of the Italian reformer Peter Damian (d. 1072), who lived for many years among the hermit-monks of Fonte Avellana, contain an abundance of advice to fellow spiritual warriors. Address-ing William, a hermit who had delayed his project of joining Peter's community, he adopted the tone of a commander ordering one of his men into battle:

> Do not let the spears hurled by your enemies frighten you, nor the tempting delights of home make you weak, the noise of the war-trumpets stun you, nor the dense forest of spears keep you hidden in your room. Now, like a famous warrior (*insignis bellator*) who has conquered all his fear, throw yourself upon the enemy lines like lightning hurled from heaven. Take up arms manfully (*viriliter*), and bearing Christ's banner, rush in a rage to where the battle seems most desperate. Make haste to smite with your sword everyone nearby, remembering always to defend yourself on every side with the shield of faith.[38]

Adding that this imagery is no 'novel imagining of [his] own', Peter claims he is merely paraphrasing Paul (presumably I Cor 9:26, Eph 6:11–17). But while Paul may have provided the basic themes, Peter has glossed the scriptural passages in interesting ways, transforming Paul's spiritual soldier into an emphatically masculine hero who is equally reminiscent of the great warriors of Classical epic.[39] The passage's emphasis on the warrior's love of combat and fearlessness also brings to mind slightly later vernacular poetry celebrating martial themes. Indeed, the following description of knightly masculinity from the twelfth-century *Chanson de Guillaume* closely parallels Peter's picture of his ideal *miles Christi*: 'No great battle can be well fought unless by the endurance of the warriors and the staying power of the young fighters, the strong and vigorous, the bold and famous.'[40]

By reading the Scriptures, *vitae* of the martyrs and Desert Fathers, a man like Peter, who came from a humble background with no familial connections to the warrior elite,[41] learned to speak the language of war with the facility of a hardened veteran. In the sermons Peter composed for his monks, spiritual warfare became a space for the enactment of heroic fantasies in which listeners were invited to imagine the feel of a sword in their hands and the hot breath of an enemy on their necks. These texts suggest Peter used preaching to initiate less experienced men into the imaginative combat in which he was himself a self-described 'old veteran', in hopes that they might in time acquire the manly *virtus* needed to fight their own battles.[42] Consider the following passage from one of his sermons for the feast of Saint Christopher:

> Still the new soldier snatched up the sword of the spirit, that is, the word of God (Eph 6:17), and then went forth into battle; and so he contended in implacable battle against the enemy of humankind…He joined hand-to-hand combat with weapons, and mowing down whatever was in his way cut down the proud

necks of the enemy. Armed with the hauberk of faith and relying on a citadel raised up not from a mountain but from the mind, although the battle-line of the host raged and gnashed its teeth on every side, the strongest athlete knew not how to fear the assault of warriors.[43]

This sort of thick description is typical of Peter's narratives of spiritual combat, which he spiced with a wealth of realistic military details that would have been instantly recognizable to his listeners, including siege engines, coats of mail, spirited horses and vats of boiling oil. Through the use of martial rhetoric in connection with heroic exemplars, Peter encouraged his monks to train meditatively for the constant battle against temptation, and offered elaborate battle scenarios into which they might insert themselves as heroes. Insisting that spiritual progress depended on constant testing of their mettle in imaginative fighting, Peter presented the *miles Christi* not only as a rhetorical device or metaphor but also as a powerful and explicitly masculine meditative tool to be employed by religious men at all stages of their careers.

The early Cistercians were fascinated by the concept of spiritual warfare and developed it at length in sermons and works of hagiography.[44] As Conrad Rudolph has shown, violent imagery in early Cistercian manuscripts operated as a kind of scribal exegesis on the ideal of spiritual warfare, and similarly complex martial allegories abound in texts produced within the Order.[45] One contributing factor may have been the Cistercian emphasis on adherence to the letter of Benedict's *Rule*, which repeatedly describes the monastic life in martial terms. This interest is also in keeping with the self-conscious Cistercian identification with the great *milites Christi* of the early Church, as well as the incorporation into their ranks of many noblemen who came to the monastic life as adults with experience of combat.[46] Martha Newman has eloquently argued that military allegories in Cistercian sermons employed familiar images of towers, moats, shields and lances to appeal to potential members, and assure young laymen that many of the manly qualities they valued in themselves and their peers – loyalty, bravery, desire for glory and adventure – need not be renounced inside the monastery.[47]

Cistercian writers likened spiritual progress to an ongoing battle between the preeminent monastic virtue of *humilitas* and *superbia*, or Pride, the root of all vices and deadliest enemy of the monk. In one of his many sermons, Bernard of Clairvaux asked each of his monks to imagine himself as David facing Goliath (a stand-in for Pride), felling the

giant with stones hurled from the sling of 'long suffering', and finishing
him off in a bloody fashion just as the young hero had done:

> Now come up closer, lest perhaps he rise up again; and, standing
> over him, cut off his head with his own sword's point, destroy-
> ing vainglory with the very vainglory that assails you. You have
> slain (*peremisti*) Goliath with the sword of Goliath if, struck by that
> haughty thought, from it you take the material and occasion for
> humility....[48]

The martial fantasy Bernard offered here (to an audience of men who
had laid aside weapons forever, but who in many cases undoubtedly
possessed personal memories of combat or military training) must have
particularly appealed to him, since he also employed it to describe his
rhetorical 'combat' with the great teacher Abelard.[49] But Bernard encouraged each listener to experience the bloody encounter for himself in
meditation, and repeatedly used the second-person singular to remind
the individual monk that this was not simply an ancient tale of combat but a description of a violent struggle even now underway within
his own soul. Here, Bernard's spiritual knight is armed only with a
sword, but elsewhere the holy abbot updated the concept of the monastic militia through the addition of recognizably contemporary elements,
inviting monks to imagine themselves mounted on chargers, outfitted
in full knightly panoply of armor and weapons, all the components of
which were tagged with symbolic meanings appropriate to the monastic
ascesis.[50] To appreciate the highly personal associations of such imagery
in the minds of Bernard's monks, we must remember that many who
entered the cloister as adults publicly renounced their worldly position
by laying their arms or armor upon the altar of the monastic church.[51]
This ritual underscored the degree to which such martial accoutrements
defined the social identity of male warriors, and might have been experienced by some initiates as a symbolic emasculation. A sermon on
spiritual combat might be an abbot's way of assuring his monks that
they had merely exchanged profane worldly arms for more powerful
spiritual weapons, and encouraging them to understand their decision
in terms that eased the transition into their new masculine identity.

If each monk was an armored *miles*, his monastic superior was his
courageous general, standing in the place of Christ, the supreme commander of the heavenly armies, and his brethren were fellow-soldiers
who shared in his victories. Abbots led their troops in battle as long
as they lived, and joined the saints in prayer for their soldiers' victories

after their deaths. In a eulogy for his predecessor Robert of Molesme, the second abbot of Cîteaux Stephen Harding (d. 1134) remembered Robert as a father-figure with whom he had shared a close martial camaraderie, lamenting that he had lost 'not only a father and ruler, but a companion and fellow-soldier (*commilito*), and a remarkable athlete in God's war'.[52] It was the duty of abbots and *seniores* to supervise the spiritual development of novices, who entered the monastery as raw recruits whose inexperience in combat left them vulnerable to demonic attack. Anselm of Canterbury warned Lanzo, a young Cluniac novice, that the devil would seek to 'annihilate Christ's new recruit by inflicting a wound of ill will', making him regret his decision to enter the monastery, or planting the seed of wanderlust in him in the hope of making him break his vow of stability.[53] In a letter to his nephew Marinus, Peter Damian similarly advised the young monk to take his martial training seriously:

> A raw recruit (*rudis tiro*) is easily struck down in his first battle, unless he first consults the drillmaster (*campidoctor*) and is carefully trained. You, who were recently sworn into the heavenly militia, who, through the profession of this holy calling, enrolled in the junior squadrons (*pueriles alae*), know that you must eagerly receive your early training among those spiritual camps from worthy instructors ....[54]

In older, more experienced *milites Christi*, new recruits found role models to guide them through their first campaigns and indoctrinate them into the model of masculinity that would define them for the remainder of their lives. The Benedictine Adam of Eynsham told of how, when the young Hugh of Lincoln sought entrance into the Carthusian Order at the Grande Chartreuse, his resolve was tested by an older monk who warned that the brothers were 'harder (*duriores*) than the stones themselves, and have compassion neither on themselves nor on those who dwell with them'.[55] Years later, however, when Hugh – by now a 'very strong athlete (*athlete fortissimo*)' was battling his own carnal impulses, his original mentor, the late Prior Basil, returned in the spirit to help Hugh secure a victory.[56] The novitiate also provided opportunities for martial competition among young monks, as was the case for Aelred of Rievaulx, who, 'during the time of his training (*tirocinio*) in Christ ... excelled all his comrades and fellow-soldiers (*commilitones*) in humility'.[57] The process by which older monks initiated their young charges into the culture of spiritual warfare has obvious parallels in the contemporary military training of lay adolescents,[58] and is even described using what might be termed 'chivalric' vocabulary; for instance, the noun *tiro*, connoting a knight-in-training or an untried *miles*, and the related term *tirocinium*, meaning knightly service, chivalry or even a tournament, are commonly used to describe novices training for spiritual warfare.[59] In the monastery as in the noble household, masculine identity formation proceeded along similar lines: a combination of instruction by older men and peer competition taught youths or new converts to fight and wield weapons, whether spiritual or real, and tales of heroic warriors and epic battles instilled in them an appreciation for the skills and virtues that adult men within their social group were supposed to possess.

The imaginations of their inhabitants transformed monasteries into military encampments, modeled on the Scriptural *castra Dei* full of angelic warriors (Gen 32:2). Peter Damian described how at Fonte Avellana 'monks' cells stand like the rows of tents in encampments (*tentoria castrorum*), like the towers of Sion and the ramparts of Jerusalem erected against the Assyrians.'[60] Writing to a Cluniac monk who had become a recluse, the abbot Peter the Venerable (d. 1156) reminded him that not only solitaries fought in the spiritual militia: Cluny itself constituted 'a great multitude of camps ordered for battle', in which 'none presumes to enter that battle alone, but places his trust in the right hands of his fellow-fighters (*compugnantes*).'[61] The individual monk's soul was for Bernard of Clairvaux a fortress, the various elements of which he described with great precision: at the center was a castle, surmounted by a strong tower and surrounded by inner and outer walls, around which enemies swarmed on all sides.[62] In a similar allegorical turn, Aelred of Rievaulx exhorted his monks to 'make ready a spiritual castle' within themselves, complete with moat, wall and tower, in order that Christ might dwell there as he had within the *castellum* of the Virgin Mary.[63] With its connotations of unyielding strength and sealed-off invulnerability, the castle is a significant choice of metaphor for the male monastic community, and by extension, body; purged of internal enemies and tirelessly policed from within, the monk's body (or body of monks) was an impregnable citadel,[64] the very antithesis of contemporary conceptions of feminine bodies as soft, permeable and difficult to control.[65]

Siege imagery was also applied to descriptions of Paradise, in which the heavenly Jerusalem became a city under attack by hordes of monastic warriors, all determined to batter down the heavenly gates and take their salvation by force.[66] Many such allegories begin with exegesis of Matthew 11:12 ('Ever since the days of John the Baptist the kingdom of heaven has been suffering force and the violent are carrying it away'), a common reading for the feasts of the Baptist, who was regarded by monks as a model spiritual warrior. The Cistercian abbot Guerric of Igny (d. 1157) asked his monks to imagine themselves as a cohort following John the Baptist into heavenly battle:

> Gird yourselves, I say, men of courage; and follow the leader and master of this happy militia – I speak of John the Baptist – from whose days heaven begins to be open to assault (*expugnabile*)...Follow, I say, that leader, whose banners are red with his own blood, whose deeds and triumphs you have chanted today with proper veneration.[67]

In a letter to Fulk, an apostate canon, Bernard of Clairvaux warned that Fulk's 'fellow-soldiers' would conquer the kingdom of heaven without him unless he returned to his community. 'The Lord himself is present as our supporter and guardian', Bernard assured him, 'who teaches our hands to fight and our fingers to make war (Ps 143:1).' Most importantly, if Christ, the commander leading the siege of his own citadel, did not recognize Fulk in the heat of the battle, he would not remember him on the Last Day.[68] Here as elsewhere, the ethos of spiritual warfare demanded that Christ be presented in masculine terms, as a powerful general or champion leading the monastic militia and modeling martial *virtus* for his soldiers. Just as Christ might be invoked in a maternal guise by writers wishing to emphasize his 'feminine' capacity for nurturing and unconditional love, and promote these virtues within the monastery,[69] casting Christ as a virile warrior enabled these same writers to promote 'masculine' qualities, such as fortitude and bravery, which were no less essential to the monk.

III   Religious women as *Milites Christi*

Any consideration of the relationship between martial rhetoric and monastic masculinity must also consider the significance of spiritual warfare for contemporary monastic women. Recent scholarship on religious women has emphasized the similar self-definition of monastics of both sexes,[70] and indeed within the monastic context the values most frequently associated with the *miles Christi*, such as obedience, cooperation and vigilance against temptation were *not* gendered per se. Like their male counterparts, religious women inherited a tradition of the contemplative life as service in the heavenly militia, and participated in many of the same activities, such as private prayer and *lectio divina*, that monks described in martial terms. In the syneisactic communities that proliferated in the eleventh and twelfth centuries, consecrated men and women 'fought' side by side, as at Fontevrauld, where Robert of Arbrissel (d. 1116) 'founded great armies of Christians (*agmina Christianorum*)', in which both sexes became 'Christ's recruits (*tirunculi*)'.[71] At the same time, we have seen that spiritual warfare played a particularly important role in the self-definition of new Orders like the Cistercians who were reluctant to admit women to full membership. Furthermore, in a period when it was becoming common for monks to be ordained priests, the performance of the mass was cast as a battle in which the celebrant conquered lurking evil forces, and religious women were banned from taking part in this sort of sacramental combat.[72]

Male clerics did employ martial rhetoric in texts written for and about women, and described abbesses, nuns and anchoresses as *milites Christi*. Like male hermits, the female solitaries who figured prominently in the eremitic revival of this period were perceived as warriors by male admirers. The twelfth-century Saint Albans Psalter, created by Benedictine monks of that house for the recluse Christina of Markyate, combines textual and iconographical references to spiritual warfare in its gloss on Psalm 1, which encourages the reader to 'reflect on that war' corresponding to the monastic pursuit of perfection.[73] An even more elaborate martial exhortation appears in the *Liber Confortatorius* written by the monk Goscelin of Saint-Bertin as a spiritual guide for the young anchoress Eva in the 1180s.[74] Throughout the second book of the *Liber*, Goscelin reminded his female subject to envision the religious life as a battle, offering her both male and female *milites Christi*, including the martyrs Blandina and Perpetua, as exemplars.[75] Nonetheless, Goscelin repeatedly admonished Eva that she would need to cultivate masculine qualities: recruits to the life she had chosen must be able to 'act manfully (*viriliter agite*)', be 'full of ardor, and able to wield spiritual as well as material weapons'.[76] Goscelin by no means denied holy women the opportunity to fight – indeed, his devotion of fully a third of his treatise to spiritual warfare indicates he viewed it as a particularly suitable theme for Eva's meditations – but follows Christian tradition by insisting women must possess virile strength if they hope to be successful in combat.[77]

The spiritual power of abbesses could, like that of abbots, be construed in martial terms. The Benedictine monk Guibert of Gembloux (d. 1213), a great admirer of Hildegard of Bingen, described his pleasure in visiting her, 'the commander of such a great troop (*dux tanti agminis*)', and her nuns at Rupertsberg, where one might behold 'the happy sight of the weaker sex triumphing over itself, the world, and the devil with the help of Christ'.[78] While this image of the holy abbess was unquestionably powerful, Guibert implied that unlike their male counterparts, female *milites Christi* must fight not only temptation but their very nature. In the cloister, as in the secular sphere, a woman warrior was seen as a powerful but unnatural figure. Many twelfth-century monks would likely have agreed with Gregory of Tours' assessment that, while they and their brethren fought 'as they should', women religious might do so only by laying aside their feminine nature and, to paraphrase Paul, girding themselves with the armor of masculinity.[79] However, even these few examples suggest that, just as some monastic men were more interested in spiritual collaboration with women than others, there existed a range of male clerical opinions on the subject of women as *milites Christi*.

Like their male counterparts, religious women employed martial rhetoric to position themselves within heroic spiritual genealogies. On the mortuary roll of Matilda, daughter of William the Conqueror and first abbess of La Trinité, Caen, her nuns remembered how 'in the amphitheater of this world she fought (*depugnare*) against vices as if with beasts, and was wounded'.[80] The extensive correspondence of Hildegard of Bingen (d. 1179) also includes many references to spiritual warfare. In a defiant missive to the canons of Mainz, who had placed her community under an interdict, Hildegard styled herself 'a female warrior battling against injustice', while in another letter she encouraged a fellow abbess to arm herself with spiritual weapons to fight against vice 'by strenuous war like a stalwart knight'.[81] Hildegard clearly knew how to speak as a *miles Christi*, and many of her letters to churchmen contain allusions to spiritual arms, combat and siege imagery. Only once, however, did a man (the aforementioned Guibert of Gembloux) address her in similar terms, and then it was to ask her to pray for *his* spiritual victories.[82] While Hildegard often addressed male clerics as *milites*,[83] it seems they were unwilling or unable to see her in the role they so eagerly claimed for themselves. The language of war was one most monks chose to speak only with one another; women, even the holiest ones, were not permitted to join the conversation.

IV   Conclusion

How, finally, can we explain the appeal of spiritual warfare to monastic men? To suggest that because such martial imagery is commonplace in medieval writing individual authors employed it mechanically, unthinkingly, seems to me untenable; recent work has demonstrated that even 'stock' metaphors used in hagiography and related medieval genres were freighted with special, often highly personal meanings for different individuals and groups.[84] The fact remains that, while Scripture, patristic texts, and the Classical tradition supplied monastic writers with a dazzlingly rich fund of symbolic imagery of all sorts, many writers selectively employed military allegories that allowed them to describe the monastic ideal in decidedly martial, masculine terms. It might equally be objected that spiritual combat was not *real*, but merely a collection of metaphors for clever writers to manipulate, but this view fails to apprehend how this most important of all wars was experienced by twelfth-century monks. As we have seen, writers went to great lengths to craft evocative martial narratives to serve as contemplative aids for readers or listeners, and for daily practitioners of this sort of meditation the meaning of spiritual combat must have surpassed the merely metaphorical.

For medieval monks, war was an excellent tool to think with, in part because imagining themselves as soldiers wielding weapons and engaging in combat helped them better understand the monastic life and more effectively convey its meaning and purpose to new initiates. As medieval writers undoubtedly recognized, the monastic life and the military life made similar demands: both demanded endurance, unquestioning obedience to superiors, and complete loyalty to the group, and placed great emphasis on carefully orchestrated concerted action. Representing his monks as a military cohort in a sermon might help an abbot reinforce the importance of community, reminding his sons that all must strive as one toward a common goal. In symbolically adopting the role of military commander, abbots reinforced the need for strict obedience on the part of their spiritual sons, while likening the monk's progress to a raw recruit's transformation into a seasoned veteran underscored the strenuous, lifelong nature of spiritual development.

While the rhetoric of spiritual warfare comprised a specialized language of the male monastic subculture, it was also constitutive of this subculture, insofar as it helped shape a monastic masculinity unique to its members. Recent studies have emphasized the degree to which monks conceived of the preservation of their chastity as a battle,[85] but in fact many aspects of monastic spirituality and all of the individual and communal activities that defined the religious life were perceived of in military terms. Drawing on the ancient theme of the *miles Christi* as well as the martially inflected masculinity of contemporary elites, twelfth-century monks carved out a rhetorical space in which to speak as men and perform their virility. In these rhetorical performances monks displayed their martial prowess for the benefit of fellow spiritual warriors, who were the primary audience of the texts considered here; on the whole, they seem to have been far less inclined to speak as soldiers to laymen or women, even religious women. Indeed, the insistence that monastic life *was* a war, one that could only be won with the aid of qualities coded as masculine in medieval society, promoted an explicitly gendered vision of the monastic ideal that placed religious women at a disadvantage.

Adopting the pose of a *miles Christi* as he made his profession, disciplined his body, or performed the liturgy, a monk reminded himself and his peers that, while they may have obeyed God's command by 'putting aside their lances and swords (Isaiah 2:4)', they had embarked on a lifelong battle with sin fought with spiritual arms. To the extent that the possession of martial skills and participation in war marked males as virile in medieval culture, monastic men remained every bit as manly or even more so – at least in their own eyes – as laymen. While military prowess may have constituted only one aspect of the masculinity of medieval laymen, [86] for monks who could not prove their manliness by amassing power or resources, or through sexual relationships with women, the rhetorical and contemplative practice of warfare may have taken on added significance as a marker of gender identity. But monastic writers chose warfare from a plethora of common metaphors for their vocation – when they might have instead described themselves as bees in a hive, sailors in a ship or workers in a field[87] – not because they were knights trapped in monks' bodies, clinging to worldly values or striving to emulate a knightly masculinity, but because becoming a spiritual warrior was a way of more fully becoming a monk. The fact that some virtues were prized by both worldly and spiritual warriors likely eased the transition into monastic life for men raised in families of warriors, but these men soon came to understand that the battles they would fight in future bore only a superficial resemblance to those they had fought in the world. Spiritual combat was purifying where worldly warfare could imperil one's soul; worldly knights were all too prone to vices like greed, vanity and wrath while the *miles Christi* turned humility, obedience and chastity into formidable weapons. Because he fought far deadlier enemies, the man who fought his battles within the cloister commanded a greater store of masculine *virtus*, and far more powerful allies than even the strongest secular warrior could boast.

Acknowledgments

Thanks are due to Denise Despres and Jennifer Thibodeaux, who read earlier drafts of this essay and offered numerous suggestions for improvement.

Notes

1. Bernard of Clairvaux, Letter 1, in *Sancti Bernardi Opera*, 8 vols., ed. Jean Leclercq and H. Rochais (Rome, 1957–77), vol. 7, 1–11. While this letter has traditionally been dated to c. 1119, soon after Robert's flight, Adriaan Bredero has forcefully argued for a later dating of 1124/25, and characterizes the piece not as a private communication but as an 'open letter' intended to stimulate discussion of the monastic ideal. For background on the text, see Bredero, *Bernard of Clairvaux: Between Cult and History* (Grand Rapids, MI, 1996), 218–21. For a concise discussion of Bernard's views on warfare, see G.R. Evans, *Bernard of Clairvaux* (New York, 2000), 167–71.
2. Bernard of Clairvaux, Letter 1.13, *Sancti Bernardi Opera*, vol. 7, 10–11.
3. On monastic masculinity, see Jacqueline Murray, 'Masculinizing Religious Life: Sexual Prowess, the Battle for Chastity and Monastic Identity', in *Holiness and Masculinity in the Middle Ages*, ed. P.H. Cullum and Katherine J. Lewis (Toronto, 2004), 24–42; Janet L. Nelson, 'Monks, Secular Men, and Masculinity, c. 900', in *Masculinity in Medieval Europe*, ed. Dawn M. Hadley (London, 1999), 121–42; Megan McLaughlin, 'Secular and Spiritual Fatherhood in the Eleventh Century', in *Conflicted Identities and Multiple Masculinities: Men in the Medieval West*, ed. Jacqueline Murray (New York, 1999), 25–44. On clerical masculinity more generally, see Ruth Mazo Karras, 'Thomas Aquinas's Chastity Belt: Clerical Masculinity in Medieval Europe', in *Gender and Christianity in Medieval Europe*, ed. Lisa M. Bitel and Felice Lifshitz (Philadelphia, 2008), 52–67.
4. On monastic recruitment from the nobility, see Joseph H. Lynch, *Simoniacal Entry into Religious Life from 1000 to 1260* (Columbus, OH, 1976), esp. ch. 1 and 2.
5. Charles de Miramon argues that the conversion of adult noblemen became much more common in this period; 'Embracer l'état monastique à l'âge adulte (1050–1200): Étude sur la conversion tardive', *Annales HSS* 56 (1999): 825–49.
6. For Peter Damian's martial rhetoric, see 'Speaking the language of war', in this essay. Orderic recounts the story of his oblation in his *Ecclesiastical History*, 6 vols., ed. and trans. Marjorie Chibnall (Oxford, 1969–80), vol. 6, 550–8.
7. Karras, 'Aquinas's Chastity Belt', quoting 59. Here Karras explicitly challenges earlier arguments by Jacqueline Murray ('The Battle for Chastity') and R.N. Swanson ('Angels Incarnate: Clergy and Masculinity from Gregorian Reform to Reformation', in *Masculinity in Medieval Europe*, ed. Dawn M. Hadley [London, 1999]: 160–77) that 'the use of military metaphors drawn from secular masculinity was a way of claiming that the cleric could nevertheless live up to a secular model of manhood.'

8. Constance Brittain Bouchard, *Strong of Body, Brave and Noble: Chivalry and Society in Medieval France* (Ithaca, NY, 1998), 4–5 and n. 9–11.

9. For an overview of this changing relationship, see Ernst-Dieter Hehl, 'War, Peace, and the Christian Order', in *The New Cambridge Medieval History*, vol. IV, pt. 1: c. 1024–1198, ed. David Luscombe and Jonathan Riley-Smith (Cambridge, 2004), 185–228.

10. Katherine Allen Smith, 'Saints in Shining Armor: Martial Asceticism and Masculine Models of Sanctity, c. 1050–1250', *Speculum* 83 (2008): 572–602.

11. For an introduction to the concept of rhetorical violence and relevant bibliography, see Sidney L. Sondergard, *Sharpening Her Pen: Strategies of Rhetorical Violence by Early Modern English Women Writers* (London, 2002): 15–20 (quoting 18).

12. It should be noted that the meaning of the term *miles* had evolved in the Central Middle Ages to mean a 'knight' rather than merely a 'warrior', and that monks who styled themselves '*milites Christi*' after c. 1000 were thus defining themselves in relation to a 'knightly' identity. While monastic writers 'updated' their martial allegories with imagery borrowed from contemporary military practice, however, their basic definition of the *miles Christi* (a celibate spiritual warrior who fought the devil with asceticism and prayer) remained fairly constant across the entire medieval period.

13. See Martha G. Newman, *The Boundaries of Charity: Cistercian Culture and Ecclesiastical Reform, 1098–1180* (Stanford, 1996), 24–5; following Caroline Walker Bynum, *Jesus as Mother: Studies in the Spirituality of the High Middle Ages* (Berkeley, CA, 1982), 167.

14. On the early development of this theme, see the classic study by Adolf Harnack, *Militia Christi: Die christliche Religion und der Soldatenstand in den ersten drei Jahrhunderten* (Tübingen, 1905). An example of this tendency to focus on spiritual warfare in the contexts of the early Church, crusading movement and military orders is Johann Auer's otherwise excellent entry on '*Militia Christi*', in the *Dictionnaire de spiritualité ascétique et mystique*, 17 vols. (Paris, 1937–94), vol. 10, 1210–23.

15. The most significant Pauline passages are Rom 8:3–16, 2 Cor 10:3–4, Gal 5:16–26, Eph 6:14–17, and 2 Tim 2:3–4. On Paul as the 'instructor sacrae militiae', see Peter Damian, Letter 10, in *Die Briefe des Petrus Damiani*, 4 vols., ed. Kurt Reindel (Munich, 1983–93), vol. 1, 135.

16. Tertullian, *Ad martyras*, c. 3, excerpted in Harnack, *Militia Christi*, 104.

17. Mathew Kuefler, *The Manly Eunuch: Masculinity, Gender Ambiguity, and Christian Ideology in Late Antiquity* (Chicago, 2001), 114–15. But see the female martyr-warrior imagery in the *Passion of Saints Perpetua and Felicity* (of c. 200), c. 10, in which Perpetua has a vision of herself fighting as a gladiator in the arena, and realizes that 'it was not with wild animals that [she] would fight but with the Devil'; for the text, see *Acts of the Christian Martyrs*, H.R. Musurillo (Oxford, 1972), 106–31 (at 118). In a sermon for the feast of Perpetua and Felicity, Augustine praised the women for 'bearing in the battle the name of Christ, and in the prize of battle finding their own'; *Medieval Saints: A Reader*, ed. Mary-Ann Stouck (Peterborough, Ont., 1999), 39.

18. Kuefler, *Manly Eunuch*, 111–17 (quoting 112).

19. On this relationship, see Edward E. Malone, *The Monk and the Martyr: The Monk as the Successor of the Martyr* (Washington, D.C., 1950); David Brakke,

Demons and the Making of the Monk: Spiritual Combat in Early Christianity (Cambridge, MA, 2006), esp. ch. 2.

20. *Homily 25 on Numbers*, c. 10, excerpted in Harnack, *Militia Christi*, 99.

21. As a letter written by a mature man to a young protégé who had abandoned the monastic life, Jerome's piece offered a natural model for Bernard's Letters 1 (to Robert of Châtillon) and 2 (to Fulk of Langres), and a comparison of these texts (see above, n. 1–2 and below, n. 68) with the letter to Heliodorus reveals too many thematic and linguistic similarities for Bernard's borrowing to be accidental.

22. Jerome, Letter 14 (*Ad Heliodorum monachum*), c. 2, in *Select Letters of St. Jerome*, ed. and trans. F.A. Wright (London, 1933), 28–53 (at 30). On Heliodorus, see Kuefler, *Manly Eunuch*, 275–76. For Jerome's application of the rhetoric of spiritual warfare to a woman, see his famous Letter 22 (*Ad Eustochium*), c. 3 and 39 in *Select Letters*, 52–159 (at 58–9 and 152–3).

23. Brakke, *Demons and the Making of the Monk*, esp. ch. 8: 'Manly Woman, Female Demons, and Other Amazing Sights: Gender in Combat'.

24. Eugène Manning, ' La signification de *militare-militia-miles* dans la Règle de S. Benoît', *Revue Benedictine* 72 (1962): 135–8. Manning argues *militia* has the meaning of 'service' or 'obedience', downplaying the militaristic connotations of these passages. Although these meanings are certainly present in the text, Benedict was certainly also aware of the long genealogy of spiritual warfare described here. Gregory the Great's *vita* of Benedict also ensured that later generations of monks inherited an image of the great abbot as a spiritual warrior; see Gregory the Great, *Dialogues*, 2.2.2, 2.8.10–11, ed. Adalbert de Vogüé and trans. (into French) Paul Antin, 3 vols., *Sources chrétiennes* 251, 260, 265 (Paris, 1978–80), vol. 2, 138, 166–8.

25. *RB 1980: The Rule of Benedict in Latin and English*, Prol. 3, ed. and trans. Timothy Fry (Collegeville, MN, 1981) [hereafter RB], 157.

26. RB, Prol. 40, 165.

27. RB, 1.5, 169.

28. RB, 2.20, 175.

29. Gregory of Tours, *Liber Vitae Patrum*, c. 19 (preface), ed. B. Krusch, *Monumenta Germaniae Historica, Scriptores rerum Merovingicarum* 1, pt. 2 (Hannover, 1885), 736; trans. Edward James as *Life of the Fathers*, 2nd ed. (Liverpool, 1991), 118. John Kitchen has drawn attention to 'the shedding of femininity in order to obtain the power associated with sanctity' in works like the *Life of Monegund*, and shown how this holy power is often represented in both men's and women's *vitae* through literary imagery of spiritual combat. See his *Saints' Lives and the Rhetoric of Gender: Male and Female in Merovingian Hagiography* (Oxford, 1998), 105 and 130.

30. Two notable exceptions are Murray, 'Masculinizing Religious Life', and Karras, 'Thomas Aquinas's Chastity Belt'.

31. For the increasingly important distinction between *armis* and *inermis*, and condemnation of clerical arms-bearing in particular, see Amy G. Remensnyder, 'Pollution, Purity, and Peace: An Aspect of Social Reform Between the Late Tenth Century and 1076', in *The Peace of God: Social Violence and the Religious Response in France Around the Year 1000*, ed. Thomas Head and Richard Landes (Ithaca, NY, 1992), 280–307.

32. Miller, 'Masculinity, Reform, and Clerical Culture', quoting 28. For the implications of reform for clerical masculinity, see also the provocative essay by Jo Ann McNamara, 'The *Herrenfrage*: The Restructuring of the Gender System, 1050–1150', in *Medieval Masculinities: Regarding Men in the Middle Ages*, ed. Clare A. Lees (Minneapolis, 1994), 3–29.

33. In canons of peace councils and synods, such as that held at Toulouges in 1027, monks were accorded protection from attack only if they themselves remained unarmed; see Giovanni Domenico Mansi and Philippe Labbé, *Sacrorum conciliorum nova et amplissima collectio*, 31 vols. (Florence, 1759–98; repr. Graz, 1960–1), vol. 19, 483.

34. This is a common pun in medieval Latin. *The Letters of St. Anselm of Canterbury*, 2 vols., trans. Walter Fröhlich (Kalamazoo, MI, 1990), vol. 1, nos. 86 and 117.

35. See, for example, *Letters of St. Anselm*, vol. 1, nos. 3, 12, 37 and 99.

36. William of Saint-Thierry, Arnald of Bonneval and Geoffrey of Auxerre, *Vita Prima Sancti Bernardi*, 11.55, selections in *The Cistercian World: Monastic Writings of the Twelfth Century*, ed. and trans. Pauline Matarasso (New York, 1993), 35.

37. See, for example, the emphasis on spiritual warfare in Stephen of Lissac's *vita* of Stephen of Muret (d. 1124) founder of the Grandmontines, *Vita et miraculi Stephani de Mureto*, in *Patrologiae cursus completus, series latina*, ed. J.-P. Migne, 221 vols. (Paris, 1841–64) [hereafter PL] vol. 204, 1008–46.

38. Peter Damian, Letter 10, ed. Reindel, *Briefe*, vol. 1, 135.

39. As references throughout his letters (including this one) indicate, Peter was well versed in Roman poetry and particularly liked to quote from Virgil. On Peter's intellectual influences, see Patricia Ranft, 'The Role of the Eremitic Monk in the Development of the Medieval Intellectual Tradition', in *From Cloister to Classroom*, ed. E. Rozanne Elder (Kalamazoo, MI, 1986), 80–95.

40. Quoted in M. Bennett, 'Military Masculinity in England and Northern France, c. 1050–c. 1225', in *Masculinity in Medieval Europe*, ed. D.M. Hadley (New York, 1999), 71–88 (at 75).

41. On Peter's background, see Lester K. Little, 'The Personal Development of Peter Damian', in *Order and Innovation: Essays in Honor of Joseph R. Strayer*, ed. William C. Jordan, Bruce McNab and Theofilo F. Ruiz (Princeton, 1976), 317–41.

42. Peter Damian, Sermon 33, in *Sancti Petri Damiani sermones*, ed. Giovanni Lucchesi, *Corpus Christianorum, continuatio medieavalis* 57 (Brepols, 1983), 201: 'Haec vobis, fratres charissimi, quasi per excessum diximus, ut spirituale certamen in quo Beatus Christophorus insignis triumphator enituit, etiam per sacrae historiae mysterium doceamus.' Peter describes himself as a 'veterano et emerito militi' in Letter 72.3, ed. Reindel, *Briefe*, vol. 2, 328.

43. Sermon 33, in *Sermones*, ed. Lucchesi, 196.

44. Martha G. Newman (*Boundaries of Charity*, 29) argues that 'the Cistercians incorporated a knightly aggression into their monastic life and redirected it toward spiritual, rather than material, ends.'

45. Conrad Rudolph, *Violence and Daily Life: Reading, Art, and Polemics in the Cîteaux* Moralia in Job (Princeton, 1997), esp. ch. 4: 'The Visual Vocabulary of Violence and Daily Life'.

46. Benedicta Ward has examined the early Cistercian use of the Desert Fathers in 'The Desert Myth: Reflections on the Desert Ideal in Early Cistercian Monasticism', in *One Yet Two: Monastic Tradition East and West*, ed. M.B. Pennington (Kalamazoo, MI, 1976), 183–9. On the Cistercians' recruitment of warriors, many of whom preferred to become *conversi* rather than monks (until the Order ruled against this practice in 1188), see Constance Berman, *The Cistercian Evolution: The Invention of a Religious Order in Twelfth-Century Europe* (Philadelphia, 2000), ch. 4: 'Charters, Patrons, and Communities'.

47. Newman, *Boundaries of Charity*, 25–32.

48. Bernard of Clairvaux, Sermo in dominica IV post Pentecosten: De David et Golia, et quinque lapidibus, *Sancti Bernardi Opera*, vol. 5, 202–05 (at 205); trans. by Beverly Mayne Kienzle, Sermon for the Fourth Sunday After Pentecost, in *Sermons for the Summer Season* (Kalamazoo, MI, 1991), 115–18 (at 118). For commentary on this passage, see Jean Leclercq, 'Le thème de la jonglerie chez S. Bernard et ses contemporains', *Revue d'histoire de la spiritualité* 48 (1972), 385–99 (esp. 387 and 396).

49. Bernard depicted Abelard as Goliath in the guise of a medieval knight 'armed with all his noble warlike gear' and even accompanied by a squire. For commentary, see Andrew Taylor, 'A Second Ajax: Peter Abelard and the Violence of Dialectic', in *The Tongue of the Fathers: Gender and Ideology in Twelfth-Century Latin*, ed. David Townsend and Andrew Taylor (Philadelphia, 1998), 14–34 (at 16 and n. 6).

50. See examples of Bernard's use of symbolic weaponry in Jean Leclercq, *Monks and Love in Twelfth-Century France* (Oxford, 1979), 96–7.

51. Miramon, 'Embrasser l'état monastique', 841–3.

52. Stephen Harding, *Sermo in obitu praedecessoris sui*, PL vol. 166, 1375–6 (at 1375).

53. Anselm, Letter 29, PL vol. 158, 1093–1101 (at 1095); trans. Frölich, *Letters*, Letter 37, vol. 1, 134.

54. Peter Damian, Letter 132, ed. Reindel, *Briefe*, vol. 3, 438–52 (at 439).

55. Adam of Eynshan, *Life of St Hugh of Lincoln*, 1.7, ed. and trans. Decima L. Douie and Hugh Farmer (London, 1961), 24.

56. *Life of St Hugh*, 2.2, ed. Douie and Farmer, 50–1. For a discussion of Hugh as 'the epitome of monastic masculinity', see Jacqueline Murray, 'Masculinizing Religious Life', 36–7.

57. Walter Daniel, *Vita Ailredi abbatis Rievall*, c. 8, ed. and trans. F.M. Powicke (Oxford, 1950), 17.

58. On young men's training in war, see Bennett, 'Military Masculinity', 73–6.

59. Charles Du Cange et al., *Glossarium mediae et infimae Latinitatis*, 10 vols (repr. Graz, 1954), vol. 8, 221. See above, n. 53, 54, 55 and 57 for examples of such usage in monastic texts.

60. Peter Damian, Letter 28, ed. Reindel, *Briefe*, vol. 1, 248–78 (at 277).

61. Peter the Venerable, Letter 20 (*Ad Gislebertum*), in *The Letters of Peter the Venerable*, 2 vols., ed. Giles Constable (Cambridge, MA, 1967), vol. 1, 31. This passage reflects the emphasis on spiritual warfare as a collective activity that Barbara H. Rosenwein observed in Cluniac texts of the tenth to twelfth centuries. See 'Feudal War and Monastic Peace: Cluniac Liturgy as Ritual Aggression', *Viator* 2 (1971): 129–57.

62. Leclercq, *Monks and Love*, 95.

63. Aelred of Rievaulx, Sermon 19, c. 5, trans. Theodore Berkeley and M. Basil Pennington, *The Liturgical Sermons: The First Clairvaux Collection* (Kalamazoo, MI, 2001), 264. On this symbolism, see Abigail Wheatley, *The Idea of the Castle in Medieval England* (York, 2004), 78–89.

64. Compare with the above description of Carthusian monks as 'harder than the stones' on which their mountaintop retreat was built (n. 55). The motif of ascetic conquest of the body (which then becomes unnaturally hard) is a common hagiographic convention; see Smith, 'Saints in Shining Armor', 593–5.

65. Of course, the body-as-castle motif often appeared in connection with Marian imagery (as in n. 63 above), and so might equally be seen as 'feminine'. However, the fortress-like body of the Virgin was hardly seen as representative of ordinary women's bodies, and exhortations to make a castle of one's body were more generally associated with virginity, chastity and control over physical desires within the religious life. See John Bugge, *Virginitas: An Essay in the History of a Medieval Ideal* (The Hague, 1975), 55–6.

66. Malcolm Hebron has shown that the proliferation of castles and increase in siege warfare in the Central Middle Ages was accompanied by a heightened interest in allegorical sieges on the part of ecclesiastical writers. See *The Medieval Siege: Theme and Image in Middle English Romance* (Oxford, 1997), 142.

67. Guerric of Igny, Second Sermon for John the Baptist, c. 4, in *Sermons*, 2 vols., ed. John Morson and Hilary Costello, *Sources chrétiennes* 166, 202 (Paris, 1970–73), vol. 2, 334: 'Accingimini, inquam, viri virtutis; et sequimini ducem ac magistrum felicis huius militiae, Ioannem Baptistam loquor, a diebus cuius coelem esse coepit expugnabile.... Sequimini, inquam, ducem istum, cuius vexilla proprio rutilant sanguine, cuius hodie virtutes ac triumphos debita decantastis veneratione.' Note Guerric's use of the second-person plural as a hortatory device here.

68. Bernard of Clairvaux, Letter 2.12, *Opera Sancti Bernardi*, vol. 7, 12–22 (at 22).

69. Bynum, *Jesus as Mother*, 110–69.

70. Penelope D. Johnson, *Equal in Monastic Profession: Religious Women in Medieval France* (Chicago, 1991).

71. Baldric of Dol, *Vita B. Roberti de Abrissello*, c. 19 and 24, PL vol. 162, 1053 and 1056.

72. For mass as combat, see Honorius Augustodunensis, *Gemma animae*, 1.72 ('De pugna Christianorum spirituali'), PL vol. 172, 566 and Peter Damian, Letter 111, ed. Reindel, *Briefe*, vol. 3, 254–5.

73. For commentary on this imagery, see Kathleen Openshaw, 'Weapons in the Daily Battle: Images of the Conquest of Evil in the Early Medieval Psalter', *Art Bulletin* 75 (1993): 17–38 (esp. 34–7). The Latin text is edited Otto Pächt, C.R. Dodwell and Francis Wormald, *The St. Albans Psalter* (London, 1960), 163–4.

74. For the Latin text, see the edition by C.H. Talbot, 'The *Liber Confortatorius* of Goscelin of Saint Bertin, *Studia Anselmiana* 37', Analecta Monastica, 3rd ser. (1955): 2–117. While the *Liber* has received attention for its confessional style and erotically charged language, its elaborate discourse on spiritual warfare has drawn few comments from modern readers. The most extensive study of the text to date, *Writing the Wilton Women: Goscelin's Legend*

of Edith and Liber Confortatorius, ed. Stephanie Hollis et al. (Brepols, 2004), comments briefly on the topos of spiritual warfare at 372–4 and 393–4.

75. *Liber Confortatorius*, ed. Talbot, 47–68 (esp. 53–55 for the stories of Blandina and Perpetua).

76. *Liber Confortatorius*, ed. Talbot, 47.

77. Perhaps Goscelin's female *miles Christi* was inspired by iconography of the Psychomachia, the contest between Virtues and Vices traditionally represented as women warriors, which symbolized spiritual combat in numerous manuscripts designed for the use of both monks and nuns in this period. On twelfth-century images of the Psychomachia, see J.J.G. Alexander, 'Ideological Representation of Military Combat in Anglo-Norman Art', *Anglo-Norman Studies* 15 (1992): 1–24 (esp. 4–5).

78. Guibert of Gembloux, Letter 38 (Ad Bovon), *Guiberti Gemblacensis epistolae*, ed. Albert Derolez, *Corpus Christianorum, continuatio medievalis* 66A (Turnhout, 1989), 366–79 (at 368–9).

79. This model of the female *miles Christi* seems related to the Christian concept of the *virago* developed in the patristic era, which continued to resonate in later medieval writing; on this concept, see the Introduction to Barbara Newman, *From Virile Woman to WomanChrist: Studies in Medieval Religion and Literature* (Philadelphia, 1995).

80. *Rouleaux des morts du IXe au XVe siècle*, ed. Léopold Delisle (Paris, 1866), 181.

81. Hildegard of Bingen, Letters 23 (to the canons of Mainz) and 174r (to an abbess), in *The Letters of Hildegard of Bingen*, ed. and trans. Joseph L. Baird and Radd K. Ehrman, 3 vols. (Oxford, 1994–2004), vol. 1, 79; vol. 2, 135.

82. Hildegard, Letter 103r (to Guibert) and 104 (Guibert to Hildegard), in *Letters*, trans. Baird and Ehrman, vol. 2, 24 and 28.

83. See examples in Letters 48r, 53r, 70r, 77r, 103r, 115r, 173r, 176r, 188r, 205, and 299.

84. See Thomas Head's instructive opening comments on hagiographical tropes in 'The Marriages of Christina of Markyate', *Viator* 21 (1990): 75–101 (esp. 75–6).

85. For clerical masculinity as defined by the struggle to remain chaste, see Karras, 'Aquinas's Chastity Belt'; and Murray, 'Masculinizing Religious Life'.

86. As argued by Karras, *Boys to Men*, 25.

87. Giles Constable has recently surveyed the variety of common metaphors for the religious life in 'Medieval Latin Metaphors', *Viator* 38 (2007): 1–20.

Part II

Priestly Masculinity: Reconciling Celibacy and Sexuality

5

Saxo Grammaticus's Heroic Chastity: A Model of Clerical Celibacy and Masculinity in Medieval Scandinavia

Anthony Perron

Absalon's problems began shortly after taking up his new position as archbishop of Lund, the metropolitan center of the Danish ecclesiastical province. In 1180, as the prelate was away from his see, a popular revolt flared up.[1] When he returned to establish peace, the violence instead raged further out of control, resulting in the death of one of the knights in his retinue and sending Absalon fleeing to the 'castle that he had in Sjælland', that is, Copenhagen. On his next visit to Skåne, on the heels of King Valdemar the Great, the locals, 'mad with the most wicked rage', threw stones at the archbishop, once more forcing him to retreat and prompting him to order the closure of all the churches in the archdiocese. Only when the full force of royal and archiepiscopal arms was unleashed, killing many of the rebels was an agreement reached. The issues in 'det skånske oprør' were many. Prominent among them, however, was the archbishop's insistence on a celibate clergy. Absalon would later be celebrated by Arnold of Lübeck as one who 'blossomed with the splendor of chastity', to which he 'called others by reproving, beseeching, upbraiding'.[2] The people of Skåne were not so favorably disposed to this ascetic virtue, and it was perhaps they whom Arnold had in mind when he noted that Absalon suffered 'many grave objections' in his fight against nicolaitism. According to Saxo Grammaticus's *Gesta Danorum*, our chief source for the uprising, the 'thankless race of the Scanians' denied that episcopal power was itself a necessary part of the church (asserting that their priests were subject to the people, not the prelate) and 'decreed marriages for the *sacerdotes*'.

It is clear that the demand for priestly continence was closely tied to the exercise of episcopal power and that sacerdotal chastity was by no means a popular ideal in the archdiocese. This essay will address the problem of implementing sacerdotal chastity in Denmark and ask specifically whether Saxo Grammaticus's enormous history of the Danes from their mythic beginnings to the time of Archbishop Absalon, the most important protagonist in the work, might not have been employed in the effort. On the face of it, Saxo would appear unconcerned with the question of clerical celibacy, aside from his mention of it in the context of the Scanian revolt. Yet, in dealing with the earlier portion of Saxo's massive work, that which deals with the legendary history of ancient Denmark, I will argue that into his stories Saxo subtly (and sometimes not so subtly) wove an ideal of heroic chastity that neatly supported the 'Gregorian' agenda promoted by the ecclesiastical leadership of his day. By depicting selected warriors of antiquity as suspicious of women and the corrupting effects of sex, Saxo created a model of masculinity through which Danish priests, deprived of one of the chief manifestations of virility, could nevertheless think of themselves as 'true men'.

I   The clergy's masculinity problem

In recent years, historians have devoted more attention to the question of 'manliness' in the Middle Ages.[3] Convinced that a full understanding of gender in the medieval period must not focus solely on women, scholars have increasingly asked how people in the premodern West constructed masculinity and delineated the proper expressions of male identity. And though, as current research has consistently emphasized, conceptions of manliness were always plural, the idea of what made a biological *male* into a cultural and social *man* in the Middle Ages might be summed up by Vern Bullough's formulation: 'impregnating women, protecting dependents, and serving as provider to one's family'.[4] To this we should add, for the aristocracy at least, the use of arms and the exercise of violence.[5] In the High Middle Ages, the clergy were in theory supposed to renounce these traditional demonstrations of their manhood. Church thinkers therefore had to find ways of explaining the place of sexless and weaponless males in the gender system of their time. Historians have offered varying interpretations of these efforts. While priests must have appeared to be 'womanly men' in the eyes of many contemporaries, R.N. Swanson has contended that they actually constituted an ambiguous 'third gender', one that was characterized by 'emasculinity'.[6] What they feared, Swanson argues, was that their fragile identity would be proved false if they were discovered actually to be engaged in the stereotypically male behavior priests were not infrequently accused of. Instead of angelic beings on earth – their claim to power – they would simply be men like all others. Along rather different lines, Megan McLaughlin and Jennifer Thibodeaux see an attempt by churchmen from the eleventh through the thirteenth century to craft a specifically clerical masculinity that would compete with the prevailing conception of secular manliness, to which it formed a sort of mirror image. Priests therefore engaged in spiritual warfare (against the flesh, against the Devil), entered spiritual marriage (to their churches) and fathered spiritual children (the souls of their parishioners).[7]

Out of all the ascetic demands placed on priests, celibacy would have presented especially steep challenges to their gender identity. A facility in shedding blood was a characteristic of only some men (those in the nobility). Sex, on the contrary, was a more universal male practice, one thus embedded in ideas of masculinity cutting across divisions of class and culture. Abstinence from sex for the purposes of religious devotion was, of course, not an invention of the Gregorian era. Its roots can be found in the third and fourth century when voluntary virginity served as a reaction to the increasing worldliness of Roman Christianity.[8] The standard was once again touted by Carolingian church policy, which aimed to sharpen the distinction between clerical and lay society as evidenced in the acts of Carlomann's synod of 742, the letters of St. Boniface and Charlemagne's famous Capitulary 'De Missis'.[9] Abandoned in the crises of the tenth century, however, the desideratum of a chaste clergy was revived in the eleventh century as the clarion call of papal reformers, codified especially at successive Lateran councils.[10] Indeed, what made the requirement of celibacy in the High Middle Ages different from previous expressions of the ideal was the extent to which a legal apparatus was applied to enforce it.[11] Such a dogged determination to forbid the clergy from engaging in sexual intercourse was, in the view of Jo Ann McNamara, responsible for a 'psychological crisis' among the priesthood of the Gregorian age. Men consigned to celibacy (willingly, reluctantly or ambivalently) could no longer 'deploy the most obvious biological attributes of manhood' and thus 'came dangerously close to traditional visions of femininity'.[12] As McNamara argued, clerical males could, and did with great alacrity, prove their 'manhood' by persecuting those whose mere existence might have called it into question, especially women. Jacqueline Murray, meanwhile, has shown how sacerdotal celibacy in the twelfth and thirteenth centuries could be seen as a symbolically martial existence, in which the 'transformation of the chaste life into the "battle for chastity" allowed men to overcome an enemy and deploy the language of military prowess without wielding a weapon or shedding blood.' Continent priests acted out their masculinity by, to borrow Hugh of Lincoln's memorable phrase, 'soldiering against themselves'.[13] In what follows, I will argue that Saxo Grammaticus in effect reversed the operation, associating martial virtue with the celibate life by reading back into the murky Norse antiquity and 'discovering' that warriors of bygone days, like the priests of the author's own generation, eschewed marriage and sex precisely as weak and feminizing and prescribed continence as a means of ensuring martial excellence. Before delving into the *Gesta Danorum* more deeply, however, it is necessary to set the context by examining the efforts to introduce chastity to a rapidly changing Scandinavian church.

II 'Europeanization' and celibacy in the Nordic sphere

Nordic society from the later eleventh to the early thirteenth century was undergoing a profound transformation. Like many other 'remote parts' of Europe at this time, Scandinavia was becoming more fully a part of the Latin West. The change began with conversion to Christianity, but its manifestations went far beyond the profession of faith. As Robert Bartlett has pointed out, Christianization and the establishment the church's institutional framework 'were preconditions for the deeper cultural incorporation' of the region beginning in the twelfth century.[14] A paradigm of 'Europeanization' has accordingly emerged to explain the dynamics of this process.[15] While conversion itself introduced deep changes, Europeanization involved a still more invasive and thorough alteration. In church matters, this meant the acceptance of papal authority and an ambitious reform ethos. Institutionally, the later twelfth and early thirteenth centuries witnessed the proliferation of communities belonging to international religious orders (Cistercian monks, Augustinian and Premonstratensian canons, and, later, Dominican and Franciscan friars), and often led by foreign abbots and priors imported for the purpose.[16] Educationally, it saw elite clerics sent for training in foreign centers, most notably Paris and Bologna, men who brought back with them ideas about reform that impacted all classes of society. For the laity, the 'reformation' of the High Middle Ages meant the imposition of new standards governing marriage, tithing and penance.[17] For the religious, both secular and regular, it involved a rigorous ideal of segregation from the corrupting ambient world. Within monastic communities, this discipline resulted in a strict insistence on claustration, combined with an intolerance of apostasy.[18]

Among the clergy actively engaged in pastoral care, such rigor was reflected, above all else, in the demand for clerical celibacy. We have already noted the efforts of Archbishop Absalon to impose continence on the priests of his church. Arguably still more committed was his successor in Lund, Anders Sunesen. Saxo Grammaticus himself praised the prelate for his zeal in 'recalling from weak softness to a more honorable attitude of mind those eager for a more lascivious life and excessively given over to the forces of intemperance'.[19] A Danish chronicle remembered him as 'a flower bright with the brilliance of your birth, since you are a virgin' and as one who was 'lacking in the filth and stench of the flesh'.[20] Pope Innocent III, meanwhile, applauded Anders's efforts to impose continence upon cathedral canons who 'return to their vomit, publicly keep concubines in their homes, and show conjugal affection toward them'.[21]

As the pope's letter suggests, the attempts to compel Nordic priests to renounce sex were by no means always successful. While there is limited evidence of 'grassroots' support for the continence agenda, much more suggests that the Scandinavian clergy resisted these new and more stringent standards of sacerdotal purity.[22] The failure of celibacy mandates in Sweden, for example, was clear by 1213 when priests there still declared that papal exemptions allowed them to remain in their 'public marriages'; remarkably, Pope Innocent III appeared willing to hear them out, pending an examination of the purported documents.[23] Similar claims were echoed as late as 1237 by clergy in Norway who 'alleged long custom' for their relationships, a practice they insisted had been sanctioned by Cardinal Nicholas Breakspear on his legation to the country in 1152.[24] In Denmark, furthermore, examples of married and fornicating priests appear with regularity in the available source material. The bishop of Viborg in 1176 complained to his cathedral's Augustinian chapter about a man named Bo Kjeldsen, who had made a profession as a canon at a collegiate church in Lime but then 'fell headlong into the vomit of apostasy and the adulterous bed of his wife, wretchedly so, as his numerous offspring attest'.[25] Similarly, among the pilgrims to the grave of St. William of Æbelholt was one Regner, who had lived among the canons of the late abbot's cloister, until he left for a life as a parish priest with a son.[26] Anders Sunesen, moreover, wrote to Pope Innocent III about a perplexing case that involved a serf guilty of mutilating a priest caught *in flagrante delicto* with the serf's wife. The pope allowed Anders to duly absolve the man.[27] Presumably, the crime of incontinence committed by the priest was worse than the husband's violent assault, or perhaps the second served as the rightful punishment of the first. The persistent lack of success in eradicating concubinage in Denmark can be glimpsed, moreover, in the visits of two papal legates. Gregorius de Crescentio, cardinal-deacon of S. Teodoro, issued a decree in 1223 that 'the offspring of a priest ought not to receive his inheritance', but according to the author of the *Ribe Bishops' Chronicle*, the decree 'was not observed because relatives of the priests did not want to disinherit their kin'.[28] Just 7 years later, Cardinal Otto of S. Nicola in Carcere Tulliano issued a strongly worded statute that more directly attacked the problem of incontinence by excommunicating mistresses, depriving nicolaite priests of their offices, requiring an oath from ordinands that they would not keep 'aliquam focariam uel concubinam', and suspending prelates who tolerated such misconduct.[29] To these examples, we must add the evidence to be gleaned from petitions for papal dispensation from *defectus natalium*. While this canonical impediment could arise from many circumstances, a number of them must have originated from the sons of priests wishing to follow in their fathers' footsteps. Such, for instance, was definitely the case for one Ivar, a provost on the island of Falster, who sought such a license from Pope Honorius III in 1220. The man was said to have been born of a 'priest and a woman whom he seduced as a virgin and kept as his wife according to the custom of the region'.[30]

Opposition to clerical celibacy can of course be uncovered in virtually all corners of medieval Europe.[31] In this regard, Scandinavia was certainly not unique. Yet, it should be borne in mind that, however difficult it was to convince priests throughout Latin Christendom to give up their wives, in those lands of the West with older ecclesiastical histories, monasticism had long provided an example of ascetic continence. The Gregorian effort was therefore not the introduction of a wholly new ideal, but the extension of a standard from one group within the church (the cloistered) to another (the clergy as a whole). In Scandinavia, by contrast, the mortar had barely set in the earliest monastery walls before archbishops, popes and cardinal legates began demanding chastity of the priesthood at large. At the same time, though masculinity everywhere depended on coital performance, Jacqueline Murray has argued that 'Germanic and especially Scandinavian society... valued sexual prowess in a way quite foreign to Christianized southern Europe.'[32] In Iceland, for example, as Jenny Jochens notes, 'few prominent men lived in monogamous marriages', while Norwegian kings were, almost to a one, adulterers.[33] The Swedes, according to

Adam of Bremen, 'know no limit in their copulation with women'.[34] As for Denmark, we need only to recall Adam's assessment of King Svend Estridsen, who famously 'noted carefully and retained in memory all the things that were said to him from the scriptures, except that he could not be persuaded concerning gluttony and women, which vices are natural to those peoples'.[35]

III   Bachelor kings and perilous marriages in the *Gesta Danorum*

The task of molding the Danish clergy to the standards set by elite reformers in the twelfth and thirteenth centuries was therefore not an easy one. At this juncture, it is worth asking whether Saxo Grammaticus might have put his history to the service of clerical celibacy in Denmark. Indeed, Saxo's turgid work was self-consciously a contribution to the Europeanization of the North. The author, presumed to be a cathedral canon, stated in his preface that the aim of his labors was to give Denmark a Latin history just like those in which other nations were 'accustomed to glory'.[36] Given this agenda, interpretations of the *Gesta Danorum* have understandably focused heavily on elucidating the manifold ways in which the text tendentiously reads aspects of contemporary European culture and society back into the Nordic past.[37] Can we then find in the *Gesta Danorum* evidence for the promotion of clerical celibacy in Denmark? On the surface, it appears not. As noted above, Saxo hardly deals in a direct way with clerical sexuality, or indeed, with the religious life at all. Though he must have been in orders himself, his book was, first and foremost, a paean to the warrior life. Even in the case of his main character, Absalon, whose career provides the climax of the whole tome, the author takes great pains to depict the archbishop as a man who was more a 'pirata' and a 'parens patrie' than a pontiff and who once abandoned mass to rush into battle with the Pomeranians, asserting that it was better 'to make offerings to God with arms than with prayers'.[38] This in marked contrast to his predecessor, Archbishop Eskil, often scorned by Saxo for his lacking patriotism, preference for the monastic life and excessive devotion to religious duty.[39] Yet, given that the work was patronized by both Absalon and Anders Sunesen, the two men most responsible for leading the charge against concubinage in the North, we might well expect the *Gesta Danorum* to have something to say on the matter.[40] And indeed, upon closer inspection of many of the key stories in the text, we begin to see something that might hint at a deeper purpose. Just as Saxo tended to 'martialize' his protagonist

Absalon in the later books of the *Gesta Danorum*, so too in the early books did he deftly 'clericalize' his warriors by suggesting that true manliness, though often expressed in feats of arms, was ultimately rooted in abstinence from sex and other pleasures.

One way in which Saxo makes a subtle argument concerning the virtue of the celibate life is through the recurrent character of the chaste king. Already in the first book of his work, we read about King Hadding, who, 'having reached the highest fulfillment of manly (*uirilis*) age' was said to have 'strived for constant training in arms with all yearning for desire left aside, mindful that he, born of a warlike father, ought to spend the whole time of his life on excellent works of soldiery'.[41] Such continence was, however, displeasing to his nursemaid, the giantess Hardgrep, who 'endeavored to soften (*emollire*) his brave spirit with the enticements of her love'. 'Why', she beseeches him, 'do you waste your days as a bachelor (*celebs*), pursuing arms, thirsting for blood...? Let this hateful rigor give way, let that pious heat arrive, and join the bond of love with me.'[42] Hadding's son Frode likewise was said to have kept himself from *uoluptatibus*, once he had reached the 'fullness of his virtues (*complementa uirtutum*)', in contrast to King Helge Hundingsbane, who 'matched savagery with excess' and who 'was cast so far into sexual desire that he left it unclear whether he burned more with tyranny or lust'.[43] The Swedish king Regner, meanwhile, similarly viewed it as 'disgraceful' (*deforme*) to begin his military career with marriage, fearing his prospective wife Svanhvide as one who 'tried to soften (*emollire* again) the stout strength of men (*solidum uirorum robus*) in her womanly way (*muliebriter*) and drench his unconquered powers (*uires*) with effeminate fear (*effeminato pauore*)'.[44] The famous Frode Fredegod also stood firm against his retainers' insistence that he marry, perhaps because his foster-brothers had 'polluted their nature' through *flagitia* (shameful crimes), namely rape and adultery.[45] Later, Saxo notes that King Fridlev was reluctant to marry because of the example of his father, to whom 'the impudence of [his] wife had attached serious disgrace.'[46] Of course, such purity might bring taunts from a king's opponents, as happened when King Halfdan Bjergram was mocked by his Swedish rival Sigvald for his continence. Halfdan responded by killing Sigvald and honorably defeating the berserk Hardben, whose crimes included the rape of princesses.[47]

It should be pointed out that most of Saxo's 'bachelor kings' eventually do marry and procreate; otherwise, how to explain the succession of kings from antiquity to the present? While such unions and the families that result from them were necessary, they were nevertheless unwelcome, as Saxo makes evident in his accounts of dangerous wives and daughters. King Hadding's eventual wife, for instance, stands in his way when he wants to take up the warrior life again, while his daughter manipulates her husband into plotting Hadding's assassination.[48] When Frode Fredegod was finally convinced to wed, his new wife Hanunde brought with her retainers who further contributed to the moral decay of Denmark; their *lascivia* encompassed torture, drunkenness and the sexual assault of virgins. Frode himself was rendered 'effeminate' by the rampant degeneracy of his court. Utterly lacking in martial virtue, he was despised by his people. Only by the introduction of the courageous Erik 'the Eloquent' as his chief knight, and the subsequent repudiation of Hanunde, was Frode able to build a vast empire and lay down his famous and stringent laws.[49] On other occasions in the *Gesta Danorum*, Saxo suggests that marriage ruins the virtue of good soldiers and the fraternity among them.[50] At the beginning of Book VII, we thus read that a breakdown in security under King Oluf was the result of the marriages of those sailors responsible for maritime watches, not to mention the mischievous wives of Oluf's sons Harald and Frode.[51]

In these examples, Saxo depicts sexual contact with women as destructive to the strength of warrior kings and threatening to rightly ordered political society. The gendered quality of his language is clear and follows familiar lines set out by other writers of the 'twelfth-century renaissance' (who were themselves drawing on traditions dating back as far as Isidore of Seville). Thus the association of woman with softness was stressed by Peter Comestor who said that Mary was 'not properly called a *mulier*, as it were *mollier* (softer), that is, having suffered 'softening' (*mollitiem*)'.[52] Rupert of Deutz had earlier made the same association, adding that 'he is a man (*vir*) who tends to virtue (*ad virtutem*)', and averred that 'a person of either sex is judged to be a woman (*mulier*), that is soft (*mollis*) who tends to wickedness.'[53] Gerhoh of Reichersberg similarly noted that the word 'man' (*vir*) 'comes from *virtus*', as did Aelred of Rievaulx.[54] As these words are reiterated time and again in Saxo and their meanings unfold, the battle between 'softness' and 'virtue' becomes also one between 'womanliness' and 'manliness'. Sexual contact breaks down the barriers dividing them and allows, in a sense, women to conquer men. One can then understand more fully why celibacy was an attractive expedient for maintaining masculinity and its attendant powers.

In portraying sexuality as dangerous, finally, Saxo particularly singles out the pagan gods (treated euhemeristically in the *Gesta Danorum*) for their frequently embarrassing relations with women. Odin (who 'was reckoned by the false title of divinity throughout Europe') was often made the object of ridicule because of his love for women. His vain wife Frigg, for example, had the gold taken from a statue of her husband, valuing the *nitor* of her own *cultus* above her husband's honor. What is more, Frigg 'submitted herself to illicit sex' with a member of her household and 'turned the gold consecrated to public worship into a tool of private luxury'. In shame, Odin was forced into exile.[55] It is possible here that Saxo was making a direct allusion to the dangers of clerical marriage. Odin was a 'priest' whose wife appropriated the wealth intended for the 'church', precisely one of the problems sacerdotal celibacy was intended to prevent. Odin once more appears as a wayward priest in Book III of the *Gesta Danorum*, where his love of a Russian princess leads him into *ignominia* and *deformitas*. In order to win access to her, he first dresses as an artisan (a disreputable profession for Saxo) and then, literally shedding his masculine identity, a serving woman with healing powers. Asked to cure the woman once she falls ill, Odin instead rapes her.[56] In the ensuing scandal, the other gods kick Odin out of the pantheon of Norse deities, a sort of heathen defrocking. After a time, adding simony to his crime of nicolaitism, as it were, Odin buys back his old office.[57]

IV   Celibate heroes

Perusing the above examples, one might understandably conclude that Saxo was simply arguing for caution in sexual matters. Women might be dangerous to manliness, but the proper response was to confine and duly supervise them. This is the argument made by the Swedish historian Birgit Strand in her pathbreaking study of women and men in the *Gesta Danorum*. Strand acknowledges that in the view of our author, 'all manner of lust is depicted as wholly incompatible with manly and martial virtues', but she concludes nevertheless that Saxo's morality was 'secular rather than religious' and that the *Gesta Danorum* was not advocating the separation of the sexes in keeping with the church's demands for celibacy, but simply warning men to be 'on their guard' around women.[58] As Strand observes, most of the women we meet in Saxo's work appear as wives and mothers, proper family roles in which interactions between men and women were entirely acceptable. As I have tried to show above, however, such relationships were a function of necessity and did not reflect Saxo's ideals. Kings could not but have mothers, and for political reasons they needed to have wives too. The protestations of so many of Saxo's monarchs that life would be better without these ties and the peril in which they not infrequently found themselves because of their dealings with women suggest that Saxo would indeed have preferred for men to renounce sex altogether. Only by doing so could they avoid the decadent and disorderly 'softening' introduced by contact with women and securely garrison the citadel of their masculinity.

The ideal of celibacy as a safeguard for virtue is developed more thoroughly in the *Gesta Danorum* through non-royal heroes, figures of model character who did not have dynastic or diplomatic reasons to wed. Two examples stand out in particular. The first is Starkad, a battle-hardened champion of imposing size and, in Kemp Malone's memorable phrase, 'hard primitivism'.[59] Starkad stood as a critic of all that was new and decadent in the world around him. He was especially incensed at the surrender of warriors to weak indulgence of every kind, to the detriment of their prowess and loyalty. Starkad is a reminder that withdrawal from sex was part of a broader sense of 'continence' that Saxo regarded as key to maintaining manly virtue in the face of corrupting (and womanly) pleasure. When we first encounter Starkad, in Book VI of the *Gesta Danorum*, we learn of his extraordinary abstemiousness, here specifically with respect to drink. His reserve likewise extended to plays (actors' bodily movements struck him as 'effeminate') and music (he loathed the 'tinkling of bells' as 'soft'). 'He held his spirit so far from lewdness that he could not endure even to observe it. That was how much his virtue/manliness (*uirtus*) fought against excess.'[60] Starkad's ascetic masculinity was most evident, however, when he visited King Ingjald in Denmark. There he found a court mired in luxury. The king himself tried to weaken the hero's *uirtutis rigorem*. In response, Starkad attacked Ingjald and his retainers for corroding the 'old values of continence' (*uetustos continentie mores*), once again meaning not sex in particular, but a spectrum of 'effeminate' pleasures that threatened 'manly intention' (*uirile propositum*).[61]

Elsewhere in the 'Lay of Starkad', sex is more clearly suggested among the list of sins committed at Ingjald's court. The king himself was 'softer than prostitutes', his mind distracted by concubines.[62] His wife, meanwhile, was a 'soldier of lust' and an 'impudent whore' who, 'trusting in her buttocks, rejoiced to take a penis in a shameful transgression'.[63] Locating the root of Ingjald's trouble in his marriage, Starkad advised the king to 'flee this savage bride' in order to recover the 'moral strength of [his] youth' (*probitas iuuenis*). The old warrior also provided a picture of how 'real men' behaved in the golden age he remembered from his youth.[64] Starkad thus associated masculinity with simple food and a single-minded focus on war. Soldiers of the past, he reminded the Danish court, wore harsh clothing, got by on little sleep and never shied away from tough work. How different were men of Ingjald's generation: full of lust and greed, they clouded their reason with insatiable desire and gluttony. Devoting needless energy to cultivating their fancy looks, they gorged themselves on banquets and perverted sex, 'seeking whores like rabbits seek grass'.[65] What is significant about Starkad's portrayal of genuine manliness is that while it relates to the warrior life, it is actually little concerned with battle. Of course, Starkad fought ('hating lewdness and extravagance, I pursued nothing but war', he boasts), but combat did not make him a man. Rather, he cultivated the *virtus* necessary on the battlefield by leading a disciplined continent life off it. In that sense, this model man from the recesses of the Norse oral past was able to win fame as a warrior only because he displayed the traits represented in Saxo's own day by the clergy, avoiding all that was womanly and soft, including (most significantly for our purposes here) sex.

Starkad's conviction that asceticism was proper to manliness and his association of sex and other forms of incontinence with effeminacy were reiterated later in the *Gesta Danorum* in the story of King Gorm and Thorkil the Far-Traveled. In Book VIII of Saxo's work, we read of their fantastic voyage to the distant realm of the giant Gerud, prompted by the king's insatiable curiosity about 'the nature of things'.[66] Though the mission is explicitly said to be non-military, it placed the participants in peril and required the same sort of virtuous abstinence demanded of warriors. Thorkil warns the Danes to refrain from gluttony, ill-considered speech, riches and, naturally, sex. When the expedition meets Gudmund, Gerud's brother, the giant tries to tempt the voyagers. They resist the enticement of food, but several succumb to their desire for Gudmund's daughters, whom the giant proposes to marry to Gorm's men. The consequences – for their well-being and their reputation – were grave:

> Four of the Danes accepted the offer and placed lust before safety (*saluti*). This contagion deprived the men, driven insane and out of their minds, of their prior memory of things. Indeed, after this was done, they are said to have been of little spirit (*parum animo*). If they had contained their behavior within the proper bounds of temperance, they would have equaled the fame of Hercules, surpassed the might of giants with their spirit and been renowned forever in their country for the wondrous deeds they accomplished.[67]

Worst off was the hero Bugge, himself famous for having saved 20 of his countrymen from a vicious attack by demons in Gudmund's realm. Saxo expressly cast his decadence in gendered terms as a loss of masculinity. Described as 'an insufficiently careful guardian of himself' (*parum diligens sui custos*), Bugge 'loosened the bonds of continence and, with that manliness (*virtute*) cast aside which he had thus far nurtured, he embraced one of [Gudmund's] daughters with an insatiable love and sought a marriage of his own destruction.' Like the four victims described above, Bugge too lost his memory. As Saxo concludes, 'that outstanding man, victor over so many monsters, champion of so many dangers, overcome by desire for one maiden, submitted his spirit, having wandered from continence, to the wretched yoke of lust.'[68]

Even Saxo's heroes who indulged in sexuality might nevertheless redeem themselves by forsaking women in order to display loyalty to their lord and prowess in battle, key concepts in the author's construction of masculine identity. In the famous 'Bjarkemål' ('Lay of Bjarke'), we meet the warrior Hjalte, who had 'given himself over to the embraces of a prostitute' (Saxo here uses *scortum*, a word literally meaning 'whore', but which in Saxo's misogynist lingo can simply refer to a mistress). At first, Hjalte thus cannot be said to exemplify the virtues of continence. However, as soon as he heard that his lord, King Rolf 'Krake', needed his help, Hjalte 'put bravery before luxury and preferred to seek the deadly danger of Mars rather than indulge in the gentle delights of Venus'. 'With what great love of the king', Saxo asks, 'might we reckon this soldier to have burned when he … judged it better to expose his safety to clear peril than to preserve it for lust?' Showing his contempt for women, when the 'harlot' asks whom she should marry if she should lose him, Hjalte cut off the woman's nose; he 'rendered her disfigured and punished her encouragement of libidinous hesitation with a shameful wound, thinking that her slutty mind would be checked by this loss on her face'. The warrior then goes to his friend Bjarke, who was also occupied with a woman, and urged him to abandon sex for war. Hjalte's words deserve to be quoted at some length here:

I do not command you to know sport with virgins, nor to stroke soft cheeks and give sweet kisses to young women and fondle tender breasts, nor to take pure drink or rub a smooth thigh and gaze upon snowy-white arms. I call you rather to the bitter struggles of Mars. There is a need for war, not light-hearted love. Weak softness has no

business here; the matter demands battles. Whoever values friendship with the king, let him grab his arms. For worthy spirits the lance
is most ready for battle. There must therefore be nothing timid or
light for brave men (*uiris…fortibus*). Let desire leave your spirit and
give way to arms… Let nothing equipped with luxury be near; let all
things full of rigor give attention to paying back the present destruction… One who strives for titles of praise or rewards ought not to lay
torpid with slothful fear, but go forth against brave enemies and not
fear the cold sword… Lo, his manly virtue (*virtus*) admonishes each
well-deserving man to properly follow the king.…[69]

Once more, we see Saxo's conception of gender values laid bare: manliness is a harsh business incompatible with the levity, softness, extravagance and fear associated with women. Desire for, and the company of,
women must be left aside in the pursuit of manliness/virtue. And if a
champion cannot be wholly celibate, he should at least know when to
come to his senses and brutally repudiate women and their corrupting
effects on virility.

Stories like those of Starkad, Thorkil and Hjalte, among the most elaborated in the *Gesta Danorum*, make manifest the dangers of sex to the
'spirit' of courageous men and the consequent advantages of celibacy.
In this context, it should be pointed out that the power of chastity in
generating masculinity was so strong as to be capable of turning even
celibate women into brave warriors. In a famous digression in Book VII
of the *Gesta Danorum*, Saxo comments on the 'shield maidens'. It should
not surprise anyone, he writes, that women might act as warriors and
'devote virtually all the moments of their lives to cultivating military
prowess, lest they should suffer the bonds of virtue to be loosened by
the toxin of luxury'. Such women gave up the 'softness of feminine
frivolity and compelled their womanly character to use manly sever-
ity'. Critical to their transformation was their avoidance of sex: 'These
women, as it were, forgetful of their in-born condition, and placing rigor
before enticements, yearn for war instead of kisses, and tasting blood,
not kisses, they carry out the duties of arms, not love (*armorum potius
quam amorum officia*).' Likewise, Saxo adds, these Norse Amazons put
their energies 'into killing, not into bed' (*non lecto, sed leto studentes*)
and sought out with arrows those men they might have enticed by their
appearance.[70] If chastity could turn women into men, one can more
readily appreciate its importance in preserving virility among males and
preventing them from slipping into 'feminine' softness.

V Clerical celibacy and masculinity in Europe and in the North

Although Saxo rarely mentioned the issue of sacerdotal continence in direct terms in his *Gesta Danorum*, he nevertheless laid out an argument about women, sexual activity and manliness that would have been appealing to priests of whom chastity was increasingly being demanded in the late twelfth and early thirteenth centuries. Saxo's work thus fits into a European-wide rhetoric concerning the problem of clerical masculinity. In her article on the 'Herrenfrage' of the High Middle Ages, Jo Ann McNamara has suggested that the crisis of masculinity provoked by the demand for celibacy gave rise to a misogynist image of women as 'dangerous and aggressive, poisonous and polluting'. Virgins – 'virile women' – might of course be an exception.[71] These ideas are repeated in the *Gesta Danorum*, as we have seen. Similarly, Saxo's close association of celibacy with enhanced masculinity has echoes in contemporary Latin-Christian writing. Bernard of Clairvaux, in his praise of the Templars, mocks lay knights as effete by comparison with the warriors of the 'new militia'. Their armor, he asks ironically, 'is it military insignia or womanly ornamentation?' Such warriors, the abbot claimed, care more for their hair, 'styled in the feminine fashion', than for efficiency in fighting.[72] One might of course argue that Bernard's Templars, like Saxo's virginal heroes, at least maintained their masculinity through violence. Priests, on the other hand, were forbidden that quintessentially male habit too. But the military orders, as seen by Bernard, were not compensating for some emasculating effects of celibacy by fighting. Indeed, the use of armed force is the one thing they share with womanish secular knights. What makes them more manly, by contrast, and thus distinguishes them from the effete lay *milites*, is precisely what they have in common with the clergy, their vow of celibacy.

The notion that chaste priests were more manly than those who succumbed to temptation, that chastity was in fact a proof and safeguard of virility appeared in Denmark before Saxo's work was complete. Indeed, none other than the *Gesta Danorum*'s second patron, Anders Sunesen, had expressed similar sentiments in his Hexaemeron, penned most likely while the future archbishop of Lund was teaching in Paris in the 1180s and early 1190s.[73] In his dense versified compendium of theological learning, the Danish master presages Saxo's attitudes concerning the decadent nature of women, who represent flesh and whose lust is dangerous even in otherwise licit marital intercourse. Borrowing on the same etymological topoi discussed above, Anders stated that the term 'maiden' (*virgo*) is derived from *vir* ('man'), suggesting that the proper state for a man is to avoid sexual contact and that a woman partakes of masculinity to the extent that 'the gate of shame remains untouched' (*manet integra porta pudoris*). As soon as that 'gate' is broken, Anders states, 'she is no longer a *virgo*, but is rightly called a *mulier*, having suffered, as it were, a "softening" (*mollitiem*) of the flesh.'[74] Anders proceeded to warn that priests, in order to preserve the manly courage proper to their office, must guard their virginity with care as a prophylactic against the 'softening' associated with women:

> no one should consider undertaking the pastor's work unless he has first girded up his loins, so that he who is obligated to teach pure things might not be polluted, that his vile life might not contradict the words of his mouth and make them worthless, and so that he who teaches brave things (*fortia*) might not be discovered to be weak (*mollis*).[75]

Whether the *Gesta Danorum* was directly inspired by the *Hexaemeron* or whether both works were expressions of Parisian thought in the second half of the twelfth century is uncertain. However, the similarities between Anders's rhetoric and the manner in which Saxo phrased the relationship between sexual activity and manliness are striking, especially given Anders's personal involvement in the work.[76]

VI Conclusion

Saxo's stories of chaste kings and abstinent warriors, while they represent the penetration of European norms of clerical behavior to Denmark, do much more than that. The demands of clerical celibacy were often difficult for men of the church to accept. In the North, the new standards represented not only a contradiction of traditional Scandinavian conceptions of manhood, but also an imposition of foreign ideas. Saxo solved these problems by allowing continent Danish priests to associate themselves with a heroic and specifically native history. Accepting his portrayal of the distant past, the *Gesta Danorum*'s clerical audience would have seen themselves as the rightful heirs of the glorious tradition of Norse warrior *virtus* and the ideals of 'manliness' that it supposedly represented. A few years ago, in an article on Abbot William of Æbelholt, Nanna Damsholt, echoing Jo Ann McNamara, posed the question 'Is a monk a man?' Her point was that the Gregorian reform had challenged traditional medieval stereotypes of masculinity, defined by 'violence and a striving for wealth and sex' and incompatible with the new standards of abstinence.[77] Damsholt did not discuss Saxo in her article, but the *Gesta Danorum* would seem to offer a clear answer. As revealed in his treatment of Denmark's legendary history, *only* a monk, or a similarly continent male, was *really* a man.

Notes

1. The episode discussed here is related in Saxo Grammaticus, *Gesta Danorum*, ed. Karsten Friis-Jensen, volume 2 (Copenhagen, 2006), 490ff. (lib. 15, ch. 4.11).
2. Arnold of Lübeck, *Chronica Slavorum*, ed. G.H. Pertz (Hannover, 1868), 78 (lib. 3, cap. 5).
3. For an overview of the trends in this scholarship, see Thelma Fenster, 'Why Men', in *Medieval Masculinities: Regarding Men in the Middle Ages*, ed. Clare A. Lees (Minneapolis and London, 1994), ix–xiii, as well as Lees's own introduction to the same volume. More recently, see D.M. Hadley's introduction to her edited volume *Masculinity in Medieval Europe* (London and New York, 1999), 1–18, and Jacqueline Murray's introduction to the collection of essays she edited, *Conflicted Identities and Multiple Masculinities: Men in the Medieval West* (New York and London, 1999), xix.
4. Vern Bullough, 'On Being a Male in the Middle Ages', in *Medieval Masculinities*, 34.
5. Following, to name but two articles, R.N. Swanson, 'Angels Incarnate: Clergy and Masculinity from Gregorian Reform to Reformation', in *Masculinity in Medieval Europe*, ed. Hadley, 168–9, and P.H. Cullom, 'Clergy, Masculinity, and Transgression in Late Medieval England' in the same volume, 182, who defines 'fighting and reproducing' as the 'two activities which most obviously characterized the ideal of masculinity.'
6. Swanson, 'Angels Incarnate', esp. 160–1 and 164–74.
7. Megan McLaughlin, 'Secular and Spiritual Fatherhood in the Eleventh Century', in *Conflicted Identities and Multiple Masculinities*, 25–44; Jennifer D. Thibodeaux, 'Man of the Church, or Man of the Village? Gender and the Parish Clergy in Medieval Normandy', *Gender and History* 18 (2006): 380–99.
8. For a terse survey of the history of sacerdotal continence as a spiritual ideal, see Paul Beaudette, 'In the World but not of It': Clerical Celibacy as a Symbol of the Medieval Church', in *Medieval Purity and Piety: Essays on Medieval Clerical Celibacy and Religious Reform*, ed. Michael Frassetto (New York and London, 1998), 23–46.
9. Carlomann had decreed in 742 'ut…nullus (sc. presbiter vel diaconus) in sua domu mulierem habitare permittat', while Boniface had around the same time launched his assault on the 'pseudo-prophets' Aldebert and Clemens, both of whom were charged, among many other offenses, with *luxoria* (Clemens was specifically said to have fathered two children by a concubine). See *Die Briefe des heiligen Bonifatius und Lullus, Monumenta Germaniae historica, Epistolae*, ed. Michael Tangl (Berlin, 1916), 101, 104–5 (esp. 56 and 57). For Charlemagne's capitulary 'de missis', see *Monumenta Germaniae historica*,

Capitularia regum Francorum, volume 1, ed. Alfred Boretius (Hannover, 1883), 95–6. Chapters 22 and 23 order canons and priests to live 'chastely', while chapter 24 is more specific, mandating that 'si quis autem presbiter sive diaconos qui post hoc in domo sua secum mulieres extra canonicam licentiam habere presumerit, honorem simul et hereditatem privetur usque ad nostram presentiam.'

10. On the tenth century, see Heinrich Fichtenau, *Living in the Tenth Century: Mentalities and Social Order*, trans. Patrick J. Geary (Chicago, 1991), 115–9. H.E.J. Cowdrey has dealt with Gregory VII's battle against nicolaitism in his *Pope Gregory VII, 1073–1085* (Oxford, 1998), 550–3, where he notes that Gregory was something of a moderate compared with figures like Peter Damian; see also Cowdrey's essay 'Pope Gregory VII and the Chastity of the Clergy', in *Medieval Purity and Piety*, ed. Frassetto, 269–302. In the same volume, see Uta-Renate Blumenthal, 'Pope Gregory VII and the Prohibition of nicolaitism', 239–67, which stresses the more radical nature of the Gregorian response to priestly marriage starting with Leo IX especially. Other historians have emphasized that the views advanced within papal circles in the second half of the eleventh century have roots as early as the later tenth, bound especially with the reforms of Cluny. See, for example, R.I. Moore, 'Property, Marriage, and the Eleventh-Century Revolution: A Context for Early-Medieval Communism', in *Medieval Purity and Piety*, ed. Frassetto, 179–208 (esp. 190–2). The relevant Lateran decrees include Lat. I, cc. 7, 21; Lat. II, c. 6; Lat. III, c. 11; Lat. IV, c. 14. See Giuseppe Alberigo, ed., *Conciliorum oecumenicorum decreta* (Bologna, 1973).

11. On celibacy and law, see James A. Brundage, 'Sex and Canon Law', in *A Handbook of Medieval Sexuality*, ed. Vern L. Bullough and James A. Brundage (New York and London, 1999), 36–7 in particular. Christopher Brooke has noted that the use of law greatly changed the nature of clerical celibacy, an old ideal by the time of Gratian. See his essay 'Gregorian Reform in Action: Clerical Marriage in England, 1050–1200', *Cambridge Historical Journal* 12 (1956): 18. See also John E. Lynch, 'Marriage and Celibacy of the Clergy: The Discipline of the Western Church. A Historico-Canonical Synopsis', *The Jurist* 32 (1972): 14–38.

12. Jo Ann McNamara, 'The *Herrenfrage*: The Restructuring of the Gender System, 1050–1150', in *Medieval Masculinities*, ed. Lees, 3–29. The passages cited here come from, respectively, page 8 and page 5.

13. Jacqueline Murray, 'Masculinizing Religious Life: Sexual Prowess, the Battle for Chastity, and Monastic Identity', in *Holiness and Masculinity in the Middle Ages*, ed. P.H. Cullum and Katherine J. Lewis (Cardiff, 2004), 24–42. The passages cited come from pages 27, 30.

14. The classic treatment of this theme is Robert Bartlett, *The Making of Europe: Conquest, Colonization, and Cultural Change, 950–1350* (Princeton, 1993), 289. Bartlett's study is the classic treatment of the 'Europeanization' thesis for the peripheries of the Latin West.

15. The interest in Scandinavia's integration into Europe can be seen in two recent essay collections: Per Ingesman and Thomas Lindkvist, eds., *Norden og Europa i middelderen* (Århus, 2001); and Jonathan Adams and Katherine Holman, eds., *Scandinavia and Europe 800–1350* (Turnhout, 2004). At the risk of oversimplifying the issue, the literature on Nordic 'Europeanization'

has focused on the development of centralized monarchy, the penetration of courtly culture and the practice of crusading, besides the more strictly ecclesiastical phenomena discussed below. On the first two issues, see, among many other contributions, the essays in Nora Berend, *Christianization and the Rise of Christian Monarchy: Scandinavia, Central Europe, and Rus', 900–1200* (Cambridge, 2007); Herman Bengtsson, *Den höviska kulturen i Norden: En konsthistorisk undersökning* (Stockholm, 1999). Research on crusading has focused especially on Denmark. Consult John Lind et al., eds., *Danske korstog: Krig og mission i Østersøen*, second edition (Copenhagen, 2006); Karen Skovgaard-Petersen, *A Journey to the Promised Land: Crusading Theology in the Historia de profectione Danorum in Hierosolymam* (Copenhagen, 2001). On the specific question of Denmark's relationship to Europe, see Brian Patrick McGuire, ed., *The Birth of Identities: Denmark and Europe in the Middle Ages* (Copenhagen, 1996), and Michael H. Gelting, 'Danmark: En del af Europa', in *Middelalderens Danmark*, ed. Per Ingesman et al. (Copenhagen, 1999), 334–51.

16. The literature on the religious orders in medieval Scandinavia is now quite extensive, too much so to summarize here. For a superb recent overview, see Tore S. Nyberg, *Monasticism in North-Western Europe, 800–1200* (Aldershot, 2000).

17. On the use of the term 'reformation' to describe this era, see Giles Constable, *The Reformation of the Twelfth Century* (Cambridge, 1996).

18. See Anthony Perron, 'Fugitives from the Cloister: Law and Order in William of Æbelholt's Denmark', in *Law and Learning in the Middle Ages*, ed. Helle Vogt and Mia Münster-Swendsen (Copenhagen, 2006), 123–36; and Johnny Grandjean Gøgsig Jakobsen, 'Abbed Vilhelms idealer: Det viktorinsk-augustinske syn på klosterlivets formål og idealer i højmiddelalderens Danmark', *Kirkehistoriske samlinger* (2001): 7–36.

19. *GD*, volume 1, page 74 (pref., ch. 1.2).

20. *Vetus chronica sialandie*, in *Scriptores minores*, volume 2, 63–4.

21. *Diplomatarium danicum*, ser. 1, vol. 4, ed. Niels Skyum-Nielsen (Copenhagen, 1958), no. 87. On Anders Sunesen's cooperation with Innocent over the issue of clerical celibacy, see Torben K. Nielsen, *Cølibat og kirketugt: Studier i forholdet mellem Innocens III og Anders Sunesen* (Århus, 1993).

22. Backing for the imposition of clerical celibacy from the ranks of lay and even common people can be found in Sjælland in the 1120s, when the nobleman Peder Bodilsen and his chaplain Nothold initiated a legal action against the clergy of Sjælland, 'so that those having wives should dismiss them and those not having wives should by no means take them'. In the wake of Peder's challenge, married priests were allegedly assaulted, murdered and dispossessed. In the end, though the demands for continence were supported by local peasants, they were rejected by none other than the bishop of Roskilde on the pretext of clerical privilege. Such complaints, the bishop asserted, belonged 'in sancta synodo', not in the public assemblies ('in placitis'), where they had apparently been presented. The account is found in the *Chronicon roskildense*, in *Scriptores minores*, volume 1, page 26.

23. *Diplomatarium danicum*, ser. 1, vol. 5, ed. Niels Skyum-Nielsen (Copenhagen, 1957), no. 37. As Innocent put it, 'de presbyteris autem Suethie certum non possumus dare responsum nisi uiderimus priuilegium quod pretendunt.'

24. *Diplomatarium norvegicum*, vol 1, ed. C.C.A. Lange (Oslo, 1847), no. 19. The cardinal, it should be pointed out, did equivocate on the question of priestly marriage. See the classic work of the Norwegian historian Arne Odd Johnsen, *Studier vedrørende Kardinal Nicolaus Brekespears legasjon til Norden* (Oslo, 1945), 271–85. As Johnsen comments (279), '…er det lett nok å komme til den slutning at legaten ikke har forbudt presteekteskapet, men det er samtidig verd å merke seg at han ikke har gitt prestene en klar og utvetydig tillatelse til å gifte seg.'

25. *Diplomatarium danicum*, ser. 1, vol. 3, ed. Herluf Nielsen and C.A. Christensen (Copenhagen, 1976–7), no. 54.

26. *Sancti Willelmi Abbatis vita et miracula*, in *Vitae sanctorum Danorum*, ed. M.C. Gertz (Copenhagen, 1908), 358.

27. *Diplomatarium danicum*, ser. 1, vol. 4, no. 95. The man cut off the priest's nose and 'wounded him in the tongue, though he did not lose the ability to speak' ('ei nasum abscidit et lesit ipsum in lingua nec tamen loquelam amisit.').

28. The *Ribe Bishops' Chronicle* has been edited by Ellen Jørgensen in *Kirkehistoriske Samlinger* 6 (1933): 23–33, see especially 32.

29. *Diplomatatium danicum*, ser. 1, vol. 6, ed. Niels Skyum-Nielsen (Copenhagen, 1979), no. 110.

30. *Diplomatrium danicum*, ser. 1, vol. 5, no. 186.

31. For examples, see James A. Brundage, *Law, Sex, and Christian Society in Medieval Europe* (Chicago, 1987), 220–2.

32. Jacqueline Murray, 'Hiding Behind the Universal Man: Male Sexuality in the Middle Ages', in *A Handbook of Medieval Sexuality*, 129.

33. Jenny Jochens, *Women in Old Norse Society* (Ithaca, 1995), 31.

34. Adam of Bremen, *Gesta hammaburgensis ecclesiae pontificum*, ed. Bernhard Schmeidler (Hamburg and Leipzig, 1917), 251 (lib. 4, cap. 21).

35. Adam, *Gesta*, 164 (lib. 3, cap. 21). A scholium adds that Svend was not responsible for his sins in this regard since they were caused by a 'defect of the race' (*vitio gentis*).

36. *GD*, vol. 1, 72–4 (pref., 1.1). Saxo's identity has long been a subject of discussion, with some even asserting that Saxo was a layman, a claim now rejected by all mainstream 'Saxologists'. For a lucid argument as to the identity of the author of the *Gesta Danorum*, see Karsten Friis-Jensen, 'Was Saxo Grammaticus a Canon of Lund?' *Cahiers de l'institut du moyen age grec et latin* 58 (1989): 330–57.

37. Inge Skovgaard-Petersen has seen Saxo's creation as a work of Christian historiography along the lines of other 'national' histories like Bede, Paul the Deacon and Dudo, not to mention Geoffrey of Monmouth, though she also stressed Saxo's appropriation of 'universal history' of the sort written by Otto of Freising. See especially her book *Da tidernes herre var nær: Studier i Saxos historiesyn* (Copenhagen, 1987). Kurt Johannesson's *Saxo Grammaticus: Komposition och världsbild i* Gesta Danorum (Stockholm, 1978) understands the *Gesta Danorum* to be an elaborate and interlocking set of neo-Platonic allegories providing instruction above all in the cardinal virtues of fortitude, temperance, justice and prudence. Focusing on Saxo's treatment of women, Nanna Damsholt has argued in her work *Kvindebilledet i dansk højmiddelalder* (Valby, 1985) (see especially 109–54) that the work was a misogynist fantasy

intended to justify Christian patriarchy by depicting women as destructive. Karsten Friis-Jensen, furthermore, has studied Saxo's rephrasing of Old Norse oral narratives into classical Latin verse in his book *Saxo Grammaticus as Latin Poet: Studies in the Verse Passages of the* Gesta Danorum, *Analecta Romana Instituti Danici*, 14 (Rome, 1987). While some have stressed the unifying role of the Valdemars in the *Gesta Danorum's* account of the Danish past, Birgit Sawyer has persuasively argued that the author was instead committed to the supervision of political society by church authorities in her article 'Valdemar, Absalon, and Saxo: Historiography and Politics in Medieval Denmark', *Revue belge de philologie et d'histoire* 63 (1985): 685–705. More recently, Kurt Villads Jensen has stressed Saxo's role in justifying Danish hostility the Wends as a facet of the crusading movement in the Baltic, while Sigurd Kværndrup has analyzed the whole *Gesta Danorum* as a unique treatise on Christian pedagogy. See Kurt Villads Jensen, 'The Blue Baltic Border of Denmark in the High Middle Ages: Danes, Wends, and Saxo Grammaticus', in *Medieval Frontiers: Concepts and Practices*, ed. David Abulafia and Nora Berend (Aldershot, 2002), 173–93; Sigurd Kværndrup, *Tolv principper hos Saxo: En tolkning af Danernes Bedrifter* (Nørhaven, 1999).

38. See *GD*, vol. 2, 244–6 (lib. 14, 21.3), where Saxo praises Absalon, newly elected as bishop of Roskilde, as one who 'non minus piratam quam pontificem gessit' and 'non minus patrie parentem quam pontificem egit', and vol. 2, 524 (lib. 16, 5.1), in which Absalon, now archbishop of Lund, interrupts those 'qui sacra gestabant' and 'cupide concitatam classem obuiam hosti in altum direxit, armis, non precibus deo libamenta daturus.' As Saxo asks rhetorically, 'Quod enim sacrificii genus scelestorum nece diuine potentie iocundius existimemus.'

39. In book fourteen (*GD*, vol. 2, 418 [lib. 14, 45.3]), Saxo explains that while Absalon was fighting the Rugians, 'Eskillus, nequid consuete religionis omittere, primam sacris operam exhibebat.' Later, when describing Eskil's retreat to Clairvaux in the 1160s during a quarrel with King Valdemar, Saxo expresses disappointment that the embattled archbishop '... priuati more tranquillam uitam degens asperum patrie conuictum uoluntarie peregrinationis lenitate mutauit, iocundiorem rerum usum apud exteros quam inter ciues habiturus' (*GD*, vol. 2, 432, [lib. 14, 50.1]).

40. Saxo states that Absalon 'compelled [me] by the power of his exhortation to undertake the work'; since Absalon died while Saxo was still engaged in the project, the author then asks Anders Sunesen to lend his 'leadership and authority' to the matter (*GD*, vol. 1, 72 [pref., 1.1 and 1.2]).

41. *GD*, vol. 1, 104 (lib. 1, 6.1).

42. *GD*, vol. 1, 104–6 (lib. 1, 6.2).

43. *GD*, vol. 1, 136, 158 (lib. 2, 1.1 and lib. 2, 5.2).

44. *GD*, vol. 1, 146–8 (lib. 2, 2.6–9).

45. *GD*, vol. 1, 276–8 (lib. 5, 1.3–5).

46. *GD*, vol. 1, 370 (lib. 6, 4.1).

47. *GD*, vol. 1, 448–52 (lib. 7, 2.8–11).

48. *GD*, vol. 1, 126–34 (lib. 1, 8.18–27).

49. *GD*, vol. 1, 282–4, 310–16 (lib. 5, 1.11–14, 3.20–24).

50. Saxo's views on the potentially corrupting influence of women on male virtue can also be found in other texts from the Norse tradition. As Bjørn

Bandlien notes (specifically discussing Hávamál), 'Women cause men to forget their social obligations, companionship with other men, and therefore the lover's social position.' See his *Strategies of Passion: Love and Marriage in Old Norse Society*, trans. Betsy van der Hoek (Turnhout, 2005), 27, as well as 36–41.

51. *GD*, vol. 1, 440 (lib. 7, 1.1–2).

52. Peter Comestor, *Historia scholastica*, in PL 198: 1649, 1723. The etymology is repeated on col. 1723, as well as in Peter's sermon to nuns on the feast of the purification (PL 198: 1745).

53. See Rupert's *Vita S. Herberti*, in PL 170: 414.

54. Gerhoh of Reichersberg, *Expositionis in psalmos continuatio*, in PL 194: 978. Aelred of Rievaulx, *Sermones*, in PL 195: 406.

55. *GD*, vol. 1, 112–3 (lib. 1, 7.1–2). As Saxo described Odin, 'duplici itaque ruboris irritamento perstrictus plenum ingenui pudoris exilium carpsit eoque se contracti dedecoris sordes aboliturum putauit.'

56. On this and other examples of rape in Saxo, including those from Frode's court mentioned above, see Nanna Damsholt, 'Women in Medieval Denmark: A Study in Rape', in *Danish Medieval History: New Currents*, ed. Niels Skyum-Nielsen and Niels Lund (Copenhagen, 1981), 71–93.

57. *GD*, vol. 1, 204–10 (lib. 3, 4.1–11).

58. Birgit Strand, *Kvinnor och män i Gesta Danorum* (Göteborg, 1980), 242–7. The heart of Strand's argument is summarized on page 246: 'En lärdom man kan draga ur Saxos historia är att det forntida hjälteidealet var oförenligt med sinnlighet och njutningslystnad. Och lika främmande var dessa laster för kyrkans ideal: kyrkans renhetskrav och – inte minst – celibatskrav medförde en utrensing av många 'okyska' seder. Osedlighet är kyrkan förhatlig, och det verksammaste medlet (för männen) var att helt undvika kvinnor ('cave mulierem'). Saxo lär männen att taga sig i akt för, men knappast att helt undvika kvinnor.'

59. Kemp Malone, 'Primitivism in Saxo Grammaticus', *Journal of the History of Ideas* 19 (1958), 98.

60. *GD*, vol. 1, 384 (lib. 6, 5.8–10).

61. *GD*, vol. 1, 412–16 (lib. 6, 8.5–12). Saxo, reflecting persistent concerns in Denmark over the hegemony of the Holy Roman Empire, here and elsewhere associates such decadence with Germany.

62. *GD*, vol. 1, 420, 432 (lib. 6, 9.5, 9.15).

63. *GD*, vol. 1, 424 (lib. 6, 9.9).

64. See *GD*, vol. 1, 426 (lib. 6, 9.10–11), where Starkad declared that 'fortium crudus cibus est uirorum, nec reor lautis opus esse mansis, mens quibus belli metitatur usum pectore forti', compares the vice of splendid cuisine so popular in his own day with the simple dishes of the past, and urges that those fond of 'rich cream' should put on a 'manly frame of mind' (*mentem uirilem*).

65. *GD*, vol. 1, 438 (lib. 6, 9.20).

66. In *GD*, vol. 1, 560 (lib. 8, 14.1), Saxo notes that Gorm '…hereditarium fortitudinis spiritum scrutandis rerum nature uestigiis quam armis excolere maluit, utque alios regum ardor bellicus ita ipsum cognoscendorum mirabilium, quecumque uel experimento deprehensa uel rumore uulgata fuerant, precordialis stimulabat auiditas.' We later learn that his desire to learn about the world extends to religious matters, including life after death (p. 572 [lib.

8, 15.1]), a curiosity that leads him to send Thorkil, who was himself Christian, to investigate Utgard-Loki, to whom Gorm was devoted. When Thorkil brought back a bad report about the pagan god, Gorm is said to have died instantly from grief (p. 578 [lib. 8, 15.13]).

67. *GD*, 566 (lib. 8, 14.10).

68. *GD*, 570 (lib. 8, 14.19).

69. 'Bjarkemål' is contained in *GD*, vol. 1, 170–86 (lib. 2, 7.1–28). The passage quoted here is on 172–4.

70. *GD*, 464 (lib. 7, 6.8). The digression is discussed by Carol Clover in her classic article 'Maiden Warriors and Other Sons', *Journal of English and Germanic Philology* (1986): 35–49, and by Holmqvist-Larsen, *Møer, skjoldmøer og krigere*, 40–58.

71. McNamara, 'The *Herrenfrage*', 6–10.

72. Bernard of Clairvaux, *Liber ad Milites Templi de laude novae militiae*, ed. Jean Leclercq and Henri Rochais, in *Sancti Bernardi opera*, volume 3 (Rome, 1963), 216. See also Andrew Holt's contribution to this collection.

73. On Anders's life and his time in Paris, see the following essays in Sten Ebbesen, ed., *Anders Sunesen: Stormand, teolog, administrator, digter* (Copenhagen, 1985): Kai Hørby, 'Anders Sunesens liv' (11–25), and Birger Munk Olsen, 'Anders Sunesen og Paris' (75–97). A synopsis of the arguments over when Anders composed the *Hexaemeron* can be found in Ane Bysted, 'Om tankerne bag Anders Sunesens *Hexaemeron*', in *Anders Sunesen: Danmark og verden i 1200-tallet*, ed. Torben K. Nielsen et al. (Odense, 1998), 53–74.

74. Anders Sunesen, *Hexaemeron*, vol. 1, ed. Sten Ebbesen and Lars Bøje Mortensen (Copenhagen, 1985), 130 (ll. 1780–7). For fuller discussion of the linguistic play in this poem (*mulier* as *mollities*, *vir* as *virtus*), its background in Latin literature and its relationship to the rhetoric of the *Gesta Danorum*, see Holmqvist-Larsen, *Møer, skjoldmøer og krigere*, 49ff.

75. Anders Sunesen, *Hexaemeron*, vol. 1, 283 (ll. 6536–41).

76. Other scholars have noted similarities between the *Gesta Danorum* and this treatise. Kurt Johannesson, for example, observed that both are structured around the four cardinal virtues (*Saxo Grammaticus*, 36).

77. Nanna Damsholt, 'Er en munk en mand?' in *Ett annat 1100-tal: Individ, kollektiv, och kulturella mönster i medeltidens Danmark*, ed. Peter Carelli, Lars Hermanson and Hanne Sanders (Göteborg and Stockholm, 2004), 120–42.

6
From Boys to Priests: Adolescence, Masculinity and the Parish Clergy in Medieval Normandy

Jennifer D. Thibodeaux

In August of 1251, the Archbishop Odo Rigaldus noted in his register[1] the case of two clerics, Ferric and Walter, both secular canons at the collegiate church of Saint Mary de Sauqueville in the diocese of Rouen. Ferric, the treasurer of the church and also a parish priest, was rumored to be involved with a woman, Alice of Garennes; the archbishop was also disturbed that Ferric had been running about town, and his derelict behavior included riding around on horseback in indecent clothing (*in habitu inhonesto*). Walter, also a parish priest and canon at the same church, had been frequenting taverns and playing dice, and was rumored of serious sexual misbehavior. The archbishop disciplined both Ferric and Walter for their errant behavior, and yet both men continued to behave badly. Sixteen years after their first recorded offenses appeared in the register, Ferric and Walter had not reformed their behavior and both had deserted their positions. Their unruly behavior suggests adolescence, and yet these men were not. They were biologically older men who continued to behave like boys.[2]

The case of Ferric and Walter suggests that some clerics did not behave according to the ideals of clerical manhood. Odo Rigaldus, as a reforming archbishop, expected and demanded that clerics behave in a manner consistent with the dignity of their vocation. Why, then, were these clerics behaving in ways so contrary? Previous scholarly work on priests and gender has pointed to an extended adolescence as the key to understanding such disorderly behavior.[3] In the Middle Ages, age was a social construct that did not always correspond to biological age. While we in the modern world might celebrate one's 'sweet sixteen' or commiserate with other 'thirtysomethings', those in the Middle Ages understood aging in terms of life phases, rather than sheer numbers. At the same time, these life phases (or life-cycles) intersected gender, creating certain criteria necessary for reaching the standards of masculinity and femininity, however they may be defined.

The Church expected that its secular clergy, those who served in the parishes, would be models of holy living for the community. After all, these men held the responsibility for guiding the souls of their parish. Becoming a 'man of the Church' required a certain degree of literacy, a dedication to moral living and a commitment to clerical manhood. Pastors were leaders in their parishes, public figures to whom others looked for spiritual guidance. Priests in particular had access to the innermost secrets of their parishioner's lives through the sacrament of confession. In cases of dire circumstances, they attended to life and death, blessing souls, giving absolution and seeing their flock on to the next life. In short, the duties of a pastor implicitly required that he be neither young nor immature in behavior.

For most studies of medieval gender, understanding age is virtually impossible due to the lack of available information. If extended adolescence offers an explanation of such illicit clerical behavior, then there should be evidence of biological age intersecting with gender performance. Using prosopography sheds new light on the behavior of the priesthood in the Middle Ages, and in this case, age and gender among the secular clerics of Norman villages and towns. Previously, medieval scholars have studied the priesthood in the Middle Ages, with an interest in parish life, pastoral duties and the role of the priest within his village. Yet until recently, scholars have neglected the subject in terms of its gender implications.[4] The gender of the parish clergy, in particular, has suffered as a subject within medieval gender studies, largely due to the available sources. Parish records can be scant or nonexistent, as in the case of thirteenth-century Normandy. For the medieval diocese of Rouen, however, there is an opportunity to study gender and age among the Norman clergy by using two available sources, the register of Odo Rigaldus, archbishop of Rouen (1248–1275) and the *Polyptychum Diocesis Rothomagensis*, or the diocesan census. The archbishop's register contains daily entries over the course of 21 years that describe the behavior of the clergy, in particular the secular clergy of the diocese of Rouen. The rich detail of this source provides a unique opportunity to analyze the parish clergy of rural Normandy, despite the lack of available parish records from the thirteenth century. The register not only records reports from parishes over the state of the church and community, it can, when used critically, provide information on the community's perception of their priest's vocational performance and gendered behavior.

The diocesan census, while in many ways a flawed source, does provide corroborating information when used in conjunction with Odo's register.[5] An even cursory glance at the register gives one the indication that clerical misbehavior was a serious problem in the diocese of Rouen. Concubinage, fornication, tavern-frequenting, gambling, brawling and wearing inappropriate clothing are all common notations found regarding the behavior of the parish clergy. There are far more egregious examples of priests behaving badly in Odo's register, but the above-mentioned behavior was indicative of the problems with clerical reform in thirteenth-century Normandy. In my previous work, I have argued that a conflict of gender identity occurred among the priesthood as the Archbishop Odo Rigaldus began the process of reforming the Norman clergy. The precise details of this conflict are obscured, as for most of the clerics studied, few documents express their personal views. An examination of the clerical life-cycle, in particular gender and age, may elucidate the behavior of the Norman clergy. Taken as isolated incidents, bad behavior among Norman clerics suggests they were an unruly bunch; but when contextualized, this evidence suggests something more: that some clerics not only were behaving badly, but they were behaving, literally, like boys.

When studying the medieval life-cycle, one thing becomes glaringly apparent: the clergy do not fit into this scheme. As vowed celibates, they were left out of a life-cycle that focused on marriage and the acquisition of a household as the key to social adulthood. The unruly behavior of certain Norman clerics illustrates a problem with the adoption of a clerical masculine identity, one negotiated after the formative period of biological childhood had passed. Was the source of this gendered behavior due to an incompatibility between the clerical life-cycle and the life-cycle for secular men in medieval life? Or was this masculine expression the result of a complex interplay between honor, status, peer and community approval in the parish?

I Being young and male in the Middle Ages

The construction of youth and masculinity was dependent upon one's place in medieval society. The state of youth was not defined by biological age, but by conditions of social status. In the Middle Ages, as much as today, youth and maturity were social constructs that existed as symbols in society. Scholars in the Middle Ages typically classified age into six groups: *infantia* (birth to 7 years), *pueritia* (7 to 14), *adulescentia* (14 to 21), *juventus* (21 to 35), *virilitas* (35 to 55) and *senectus* (55 onward).[6]

While these terms in theory corroborated societal age with biological age, in practice, these 'ages of man' did not rigidly define phases of life. In particular, we see that age as a construct became increasingly blurred beginning with the *adulescentia* (adolescence) and *iuvenes* (youth) phases of life. Typically, men over the age of 35 were considered as part of the *virilitas* (literally 'manhood') phase of life, and yet as this essay will make clear, males in this biological age group had not always acquired the status of adult manhood. While certain actions and rituals signify youth for any given period of time, it is impossible to speak of universal 'rites of passage' or 'coming of age' for medieval men.[7] In the Middle Ages, the primary marker between the perceived young and old was marriage and the acquisition of a household, which included the production of heirs. Men were not the only ones who fell into this schema. Women also experienced different phases of life, although their life-cycle differed considerably from that of men. Women could achieve social adulthood earlier, as early as 14, if they married and had children, though womanhood carried a different social status and set of legal privileges than manhood did.[8]

Social class created different experiences in the life-cycle of men. The experience of being in *adulescentia* or an *iuvenes* in medieval society would have differed if one was the son of a well-established aristocratic family from the Norman Pays de Bray or the urban teenager of a Rouennais merchant family. Take, for example, the rites of passage experienced by warriors in medieval society. These men passed through a few stages of age before acquiring adult manhood. Young boys on their way to becoming knights began as squires, assisting a knight in his preparations. Typically around the age of 21, young men of the elite received their weapons and knighthood was conferred upon them. What followed was an undetermined period of youth, that involved roaming from place to place in search of fun and adventure. They traveled in groups, participating in tournaments, acquiring wealth, carousing with women and enjoying this unstable period of their lives. This period of youth only ended with marriage and the acquisition of a marital household, factors that 'settled' the youth and forced the passage to adulthood. For example, the famous Norman knight Guillaume le Marechal remained in his period of youth for 25 years, before passing to adulthood upon his marriage.[9] Thus youth could remain for an extended period of time, even though the man might age biologically. Knights frequently remained youths well into their mid-30s. Medieval nobles were not the only social group, however, to experience the social extension of youth. This social construction of age applied to other groups in medieval society, such as medieval artisans and university students.[10] Generally, while the passage to male adulthood might have involved other factors, it almost always involved marriage and the acquisition of a marital household.[11]

For most males in medieval society, marriage signified an important role and status because it conferred social adulthood. In obvious ways, marriage brought stability and maturity for men as they took on the responsibility of provider and leader of their household. Both economically and legally, they stood as guardians of their families. From a societal perspective, married men took on adult status in the community and acquired a number of legal privileges. In many medieval contexts, married men entered into societal governance in some form or the other.[12]

Unlike married men, male adolescents still had to acquire an appropriate masculine identity and this occurred in several processes. While married men took on local governance and household responsibilities, young men, be they 'adolescents' or 'youths', behaved in very contrary ways by participating in behaviors that challenged authorities and other established powers. Youth was not defined by rigid age limits, but rather 'movement, development and liminality'.[13] Rather than shouldering familial responsibility and societal governance, youths behaved with the freedom accorded their life stage. Male adolescent behavior, in particular, frequently transgressed the boundaries of social decency. Studies of male adolescence in different premodern European contexts point to common cultural features of youthful masculinity. For instance, medieval urban youths embraced excessive drinking, roamed their towns at night, engaged in acts of youthful violence, preyed upon women and in general sowed their 'wild oats' on the way to masculine adulthood.[14] Outside of towns, sons of the warrior elite also engaged in rather rambunctious and defiant behavior against their superiors. In medieval Normandy, sources confirm this same perception of youth. From the early twelfth century, Orderic Vitalis tells of Robert Curthose, son of William I, who like a 'raw youth' with 'his wrath and ambition violently inflamed' demanded his patrimony from his father.[15] Younger men also went 'running about town at night', a phrase that describes a whole host of known and unknown nocturnal mayhem. This sort of erratic behavior was also duly noted by the officials at medieval universities. In 1269, the Paris officials excommunicated those students who engaged in disorderly behavior, including roaming the town, fighting, raping women, and breaking into houses.[16] On another occasion in 1276, the papal legate excommunicated those students who carried arms and who also had banquets and gambled in church.[17] Whether rebelling against ecclesiastical authorities or secular ones, males demonstrated youthful behavior by engaging in rituals and actions that symbolized both their masculinity and their life stage. As Barbara Hanawalt has shown in her own work, the behavior of these adolescents was thought to be a serious threat to society.[18]

Above all, youths had to negotiate their masculine identities in accordance with the approval of their peers. This was an ongoing process, one that continued until marriage, when adult manhood was obtained. Until then, youths participated in behavior that was received as disorderly by their societies. Youthful actions typically were antithetical to the holiness and moral decency valued by the Church. For youths who were also clerics, this was problematic.

II   From boys to clerics

If youth was expressed in a disorderly fashion, then how do we categorize parish clerics who also behaved in such ways? There has been quite a bit of recent research examining the construction of gender among different groups of the medieval clergy. Some scholars have pointed to a 'crisis' in gender identity among clerics, and others have interpreted the gender identity of clerics to be one of 'emasculinity'.[19] Rather than viewing the clergy in binary terms, or as a 'third gender', others have preferred to link the clergy with one of the many masculinities that existed in the Middle Ages.[20]

Secular clerics, from the highest holy order of priest to the lowest minor order of porter, interacted daily with their parishioners. In the parishes of Normandy, village clerics were often drawn from the local community; while social origins are not known for all village clerics, a certain contingent was drawn from aristocratic families. The study of monastic clergy requires a different approach, for the legacy of child oblation and the nature of the cloistered life created a different set of dynamics related to age and gender. Becoming a secular cleric could be more problematic in regards to acquiring the ideals of clerical manhood as defined by the reforming Church. Secular clerics were not raised in an exclusively male religious environment in which such an identity could be inculcated. A young boy who entered the minor orders lived in a secular environment that promoted contrary modes of behavior for males. Arguably, it might have been easier for male monastics to adopt clerical manhood because they lived in an isolated environment where they could reference no other mode of masculine behavior. So long as a male in minor orders (or major) was able to witness masculine behavior in his peers, he would still have entertained notions of a different masculinity, one perhaps more compelling than the clerical one prescribed to him. The medieval secular clergy makes an interesting case study for the examination of youth and gender, for they are a group who live among secular men, interacting with them and yet whose life-cycle does not conform to that of secular society.[21]

To begin with, the clerical stages of life do not match the stages of life for the laity. Ordination to the priesthood required a candidate to pass through all seven orders, minor and major. Even the minor orders of the clergy did not always correspond to age; for example, when one was selected for appointment to a benefice, a candidate could move through all minor orders in a relatively short period of time, sometimes even a day, regardless of how they aged biologically. Boys could be tonsured as early as 7 years of age, at which point they entered clerical life. The order of porter, whose responsibility it was to lock and unlock the church doors, was the first entry into minor orders. As porter, the boy took on a pseudo-guardianship of the parish church. The next order, lector, required some degree of literacy for it allowed the youth additional responsibility by permitting him to read from the scriptures during services; literacy was one of the primary markers of clerical identity. The exorcist assisted the priest with blessings.[22] The order of acolyte, one who assisted at the altar, had a minimum age requirement of 14 years, and in terms of spiritual duty, carried the greatest responsibility of all the minor orders. The acolyte was the first stage at which a youth assumed spiritual responsibility; it was the only of the minor orders to be recorded in the ordination lists of Odo Rigaldus. While ordination to acolyte did not bring the youth clerical manhood, it nonetheless marked his first foray into the realm of spiritual adulthood, as those who legitimate access at the altar were designated the 'men' as opposed to the 'non-men'; women were historically denied legitimate access to the altar, for it was long believed by church authorities that they were sources of pollution.[23]

For a man to enter the major orders of the Church, that is, the subdeaconate, deaconate and priesthood, one would have to meet the minimum age requirement of 18, 19 and 25, respectively.[24] Perhaps more importantly, these major orders also required celibacy. A candidate's first steps into clerical manhood was as subdeacon. The subdeacon read from the scriptures during Mass and prepared the sacred vessels for communion. Not only did the subdeacon have access to the altar, which was a spiritually masculine space, but he also touched the sacred vessels that would contain the body and blood of Christ during communion. The deacon could participate in official sacramental duties, like perform baptisms and marriages. Once a candidate was ordained to the priesthood, he obtained full pastoral ministry. A priest governed over all sacraments (except confirmation and ordination). Administering the sacraments required maturity and responsibility, for errors made could jeopardize the soul. Changes in the administration of confession gave greater jurisdiction over penance to the priest.[25] Instead of assigning a penance according to a prescription, the curate had to judge the penitent's sins, something that required maturity and a sense of responsibility. The very nature of a priest's duties demanded personal stability and residence in the parish. Sacraments like baptism, confession and extreme unction required competence and, sometimes, immediate attention. The priest carried enormous spiritual duties. In sum, he was charged with the spiritual well-being of his flock. Once again, the differences between the life-cycle of secular men and clerics must be recalled. At a time when secular men were experiencing their *adulescentia* and *iuvenes*, a period of sexual promiscuity and otherwise disorderly behavior, clerics in major orders were assuming spiritual responsibility, with its important component of celibacy.

As P. H. Cullum has argued, clerical identity had to be inculcated early on in a boy's life to ensure his commitment to the pastoral life. According to her argument, clerics experienced an adolescent phase when they entered the rank of acolyte and remained in this phase until ordination to the priesthood; this 'extended' adolescence they experienced was due to the limitations on their duties as clerics, limitations that remained until their ordination to the priesthood. As clerics below the rank of priesthood, they were denied full sacramental power and authority. At the same time, many of these clerics faced conflicts of identity as they participated in behavior marked as youthful by secular society. It was not only their positions as minor clerics that marked them as socially adolescent; it was also the status of being unbeneficed. As she has observed, their period of 'youth' ended at ordination to the priesthood and the acquisition of a church benefice, when spiritual power and authority was conferred upon them. This marked the passage from youth to adulthood for clerics; as they were ordained priests, their adolescence ended and they achieved social adulthood through their acceptance of a spiritual adulthood. This spiritual adulthood entailed spiritual fatherhood for these new priests over their parish.[26]

Masculine identity, however, is a negotiated identity that depends upon multiple factors, including peer and community approval. A theoretical model formulated by a reforming Church could not have automatically conferred adult masculine status upon priests, because such an identity had to be created and sustained internally, by one's own acknowledgement and acceptance of it, and externally, by one's immediate community. It certainly *seems* true that parish clerics were in an extended adolescence when they entered the minor orders; after all, the roles of minor cleric entailed little, if any, spiritual authority, and as clerical positions they resembled most the corresponding age bracket of *adulescentia* [adolescence]. The higher orders of subdeacon, deacon and priest did, in contrast, entail spiritual authority and status; once ordained a priest, the cleric *could* experience social adulthood. But, as Derek Neal has argued, we must conceive of gender identity 'less narrowly and neatly'.[27] After all, clerical masculinity is not such a neatly packaged gender, easily adopted by every medieval man upon his ordination. We must not discount the role of peer and community approval within this gender negotiation.

As Cullum has elsewhere noted, the higher order of the priesthood should have functioned as a sort of marriage and social adulthood for the candidate.[28] Clerical manhood brought 'fatherhood' to the celibate minister and a 'manly' authority to judge and lead the spiritual 'children' of his parish. On a practical level, there existed other opportunities to develop spiritual families; within the parish priest's household, there sometimes existed *familia*, clerical employees of the rector, who served the benefice and sometimes even lived with him in his home.[29] While Cullum's argument would explain the compliant clergy, it remains the case that not every ordained man willingly accepted the gender negotiation necessary to fulfill his new role as a 'man of the Church'. It is less certain that ordination to the priesthood immediately ended this adolescence for all clerics.

A number of factors affected the acceptance of spiritual authority, but it is clear that some men successfully negotiated their new clerical identities, while others failed to do so and behaved according to their peer expectations, ones that were common to youths. A cleric on his way to the priesthood and a benefice could clearly see the benefits of such a vocation such as a steady income, a home and leadership status in the parish. But he also sacrificed much that was valued by his secular peers in medieval society.

III   Boys will be boys

The cases examined below demonstrate that peer and community approval had enormous influence over the performance of masculinity among Norman clerics. Furthermore, this gendered performance was inextricably tied to masculine bonding. Some clerics who behaved as adolescents were likely biologically younger men; yet, through a close analysis it appears that some Norman priests were actually older men who participated in immature, youthful behavior.

If, as Deborah Youngs has suggested, marriage and the acquisition of a household conferred social adulthood on medieval males,[30] then it follows that the clergy were not considered adult; this, of course, is a false construct. The Church promoted the benefice and rectory as substitutes for the household, and by suggesting that the priest married the parish, created an equivalent in the form of spiritual marriage. This was most certainly the pastoral ideal as it was envisioned by the Church. Evidence from the register of Odo Rigaldus, however, amply demonstrates that the benefice did not always confer social adulthood on the priest; furthermore, many of the beneficed and unbeneficed, both clerics and priests, behaved as youths. In fact, while in theory beneficed clerics and priests attained social maturity by 'marrying' their churches and acquiring their spiritual 'households', their behavior instead suggested men who had not achieved social adulthood but who could be classified by their society as youths.

The example of vicars and chaplains illustrates the problem with the gender performances of unbeneficed clergy. If, theoretically, the priest 'married' the parish church, would this aspect of clerical masculinity have been applied to the unbeneficed clergy, those who did not have a permanent position? It is perhaps understandable that social adulthood might be more difficult to achieve for this group. For the scores of 'adjunct' clerics, stability of career was not guaranteed. Acting as a substitute priest, not a permanent rector, meant that one did not enjoy stability of position and lifelong tenure. Thus social adulthood might have been more elusive and adolescent behavior more rampant for this group of clergy. The precarious position of the vicar, a 'substitute' for the rector, was reflected by the Fourth Lateran Council's decree to guarantee a suitable wage to vicars.[31] Chaplains were also not on stable ground in the parish either. Although priests, they served as assistants or supplementary clerics to the rector of a parish. The lesser degrees of job security might have meant less likelihood of attaining spiritual manhood, as viewed by the self and by the community. Vicars and chaplains may not have seen the production of spiritual 'children' as pertinent to their careers. The nature of their positions was such that they could easily find themselves serving another parish. Is it really appropriate to say that a priest was married to the church if one was a vicar or chaplain? To do so would liken a permanent, lifelong position to a temporary, unstable vocation. Such considerations are necessary when analyzing the gender of clerics. A so-called 'adjunct' cleric would have been in a remarkably different vocational position than a beneficed priest. How then did he negotiate his gender identity?

Looking at some examples from Odo's register provides a greater understanding of how this dynamic functioned. Vicars and chaplains had no real household acquisition, clerical or secular, nor did they achieve full spiritual adulthood while working for the rectors of a parish church. Another group of clerics engaged in pastoral care were canons, who served cathedrals and collegiate churches. In 1263, Odo's register recorded a detailed visit the archbishop made to St-Hildevert-de-Gournay, a collegiate church housing a chapter of canons. The archbishop found that the vicars and chaplains were performing their duties inadequately and neglecting daily prayer. Perhaps more disturbing, the clerics were raucously celebrating the feast of Saint Nicholas by dancing and singing throughout the streets of the town. This incident seems to echo the behavior of university students, who celebrated festivals in a similar fashion.[32] At the same visit, Odo found that a couple of chaplains, Simon and Lawrence, were playing dice. Both clerics had been accused of violent behavior. In particular, Simon had allegedly beat a woman and raped her daughter. A vicar named William was noted for beating a man in town and pulling a knife on him.[33] Three years later, on another visit to the chapter, Lawrence was named again for the particular offenses of playing dice, playing ball and drinking in taverns, to which he confessed. In addition, he was accused of impregnating a woman, and to this accusation, he fully denied any truth.[34] On yet another occasion at the same church, the archbishop noted that a group of clerics had been running around town at night, robbing Jews, beating up townsmen and fishing in the king's pond.[35] Living in collegiate chapters was in many ways an 'infantilizing' experience for clerics.[36]

The behavior noted above seems more reminiscent of a modern fraternity house than a medieval community of clerics. Homosocial bonding occurred in many contexts, and sources indicate that medieval youths frequently exercised masculine behavior in groups and, in some cases, in pairs. Masculine pair bonding occurred in numerous contexts. Ruth Karras has pointed out the homosocial bonding that occurred with 'brothers-in-arms', pairs of knights who fought together, who sometimes even formalized their relationship with contracts of their professional relationships.[37] There is significant evidence of similar masculine pair bonding among the secular clergy, both canons at collegiate churches and co-rectors of parish churches. Odo Rigaldus cited a number of clerics for bad behavior, many of them paired with a *socius*, or associate. In places where nepotism functioned, these associates may have been related by blood. Two canons, Ralph and Vincent Pont de l'Arche, were nephews of the aging and ailing bishop of Lisieux, William Pont de l'Arche, and both had received prebends at his cathedral. In January 1249, Odo undertook an investigation of his suffragan see, and found that these two men had been running about town at night carrying weapons and frequenting the brothels; they were also noted for their fornication with many women. The canons were ordered to leave town and attend the university, and to not return without the permission of the archbishop. Eight years later, both canons were noted again for similar derelict behavior. Ralph was investigated for the rape and strangulation of a townswoman, while Vincent was again suspected of involvement with a widow, who had given birth to his child.[38]

The members of the collegiate church of Saint-Hildevert-de-Gournay confirm that chaplains, vicars and other clerics in minor orders did not always assume clerical manhood and social adulthood. Accordingly, as low-paid substitute clerics, these positions did not always offer the career mobility to better paid and higher status positions. In such a context, it is not difficult to imagine that peer approval had considerably more influence over masculine behavior than clerical manhood did. It is also unlikely that their local communities would have viewed them as socially adult. But if we assume ordination to the priesthood brought maturity, along with the acquisition of a benefice, then we should not find such behavior among beneficed, parish priests. Yet, such behavior among this cohort did exist and was amply recorded in Odo's register.

The youthful behavior expressed by these priests cannot be taken as proof that they were biologically younger males. While the possibility exists that the parish clerics previously discussed were really biologically younger men, the following cases suggest otherwise. If we recall that clerics entered major orders at 18 years, then it would be expected that some new clerics (and in particular, priests) had difficulty adjusting to their new adult roles. With the acquisition of a church benefice, they acquired spiritual authority and responsibility; this new role they assumed at a time when their peers were perhaps still sowing their wild oats. Yet, a close prosopographical analysis demonstrates that some of these clerics really were biologically older males.[39] Ever cognizant of the errors in Odo's register and the census, I chose the smallest cohort of clerics for which I had confirmed identities. My analysis, while reliant on a small sample, nonetheless indicates that some Norman clerics continued to behave as youths, even after entering the major orders and acquiring benefices. Furthermore, some of these errant priests were biologically older men, who persisted in their illicit behavior for over a decade, and whose lay peers would have been married householders and leaders of their communities.

Just as we saw such behavior among the unbeneficed clergy, the same homosocial bonding occurred among the beneficed clerics. Such pairing among clerics could prove beneficial as each helped the other to establish his spiritual authority and negotiate his way into his clerical vocation. Yet, this association also proved to be detrimental as it could make reform difficult when one or both clerics refused to change their ways. Ferric and Walter, secular canons and beneficed priests, engaged in delinquent behavior together for over 16 years. The documents provide little information on the age of these men; however, since their errant behavior continued over a course of 16 years, it is clear that as they aged, they continued to exemplify behavior associated with the adolescent element of society. Our knowledge of these clerics begins with a record of Odo's visitation to the collegiate chapter of Sauqueville in August 1251. Ferric, the treasurer of the chapter, was then rumored to be involved with Alice of Garenne; he was also making a public nuisance of himself, as he was going about town and riding around on a horse wearing a hood without a cape. Odo noted that Walter was playing dice in the taverns and had also been linked to breaches of chastity, although nothing specific had been confirmed. A year later, both men still persisted in their adolescent behavior, and were seen at the tavern. Although both men were not fulfilling their duties, they still collected their share of the 'commons', the supplemental income that canons collected while in residence. In March 1254, Ferric and Walter were still going to the taverns. Ferric refused to wear a clerical cape when he attended church, and had in fact been absent from the parish for a while. After reiterating his previous warnings, Odo ordered the seizure of their church revenues. The archbishop was later successful in getting Walter to resign his portion of a parish benefice. Yet, the problems with these two clerics continued. Ferric stood trial for sexual misconduct with Albereda of Caney in 1261. Other canons at Sauqueville reportedly disliked Walter because he was a menace to their community; they also mentioned Walter's illiteracy, altogether insinuating that Walter was unsuitable for clerical life. Disciplinary action, however, had little effect on their behavior. On his last visit to Sauqueville in 1267, Odo noted that both men had deserted the church. These clerics were never reformed; eventually Odo ordered their incomes seized. [40]

Both priests were identified by their courtesy titles 'dominus', a title indicating their adult status as men of the church; yet both fully shirked their pastoral duties and behaved in a distinctly unadult, unclerical manner. These flagrant abuses encourage the assumption that Ferric and Walter felt neither guilty nor remorseful over their actions. While their individual voices are obscured by the records, both Ferric and Walter demonstrated by their behavior a complete lack of concern with the authority and discipline of the archbishop, and with their clerical vocation overall. Their rebellion against authority suggests the disorderly nature of youth. The improper clothing worn by Ferric and his frequent non-residence from the chapter indicates a desire to project a public image not associated with clerical authority and a fleeing from adult responsibilities. Odo even noted that Walter was illiterate, which furthermore indicates an incompatibility with the clerical life. Ferric and Walter could not have been biologically younger men, for they were under investigation for 16 years. Had they both met the minimum age requirement of the priestly office, 25 years, at the time of their first appearance before the archbishop, they would have been well over 40 years old by the time they were deprived of their incomes. These men were perhaps the best documented case of clerical misbehavior over time; but they were not the only such cases that appeared in Odo's register. In 1248, Ralph of the church at Bailly-en-Rivière was cited for drunkenness along with his co-rector William. Baldwin, co-rector of Saint-Rémy-en-Campagne, eventually resigned his portion of the church benefice after numerous accusations of sexual fornication; his associate (*socius*) Matthew was listed in the register as a dice-playing 'vagabond' who also stood accused of incontinence.[41]

Of course there were some biologically adult males in medieval society who rebelled against authorities, were promiscuous and exhibited violent behavior. Was the behavior noted above substantially different from that of the aristocratic *bacheliers*, journeymen, university students and urban youths of medieval society? Any sort of noted disorderly or violent behavior, frequenting the taverns, roaming the towns, rebelliousness and sexual promiscuity, *when cited all together*, could be seen as adolescent behavior. Certainly, this sort of behavior was not accorded the status and honor of social adulthood by medieval society.

These following cases allow for a more precise analysis of age and gender. Overall, these cases of beneficed priests indicate that clerical manhood was not successfully negotiated, and for that matter, neither was social adulthood. Gervaise, the priest at St-Rémy-en-Rivière, was cited by the archbishop in January 1248 for publicly known drunkenness, playing dice and frequenting taverns where he would often get into brawls. Perhaps to further complicate his public identity, he was not wearing his priest's gown, a long black robe that publicly denoted priestly status. Gervaise acknowledged the rumors of his behavior, and swore his resignation if he was found guilty of such offenses in the future. According to the census for the diocese of Rouen, Gervaise had acquired his church benefice sometime during the episcopacy of Archbishop Thibaud (1222–1229). This made Gervaise, at the time of his appearance in Odo's register, between the ages of 44 and 51. [42] In a similar case, William from the parish at Saint-Sulpice also appeared before the archbishop for his bad behavior; yet, in this case, William was cited numerous times over the course of 15 years. His first citation was for drunkenness in 1248. Odo cited him again in 1258, when he was noted for sexual misbehavior, frequenting taverns and not wearing a priest's gown. He was ordered to cease his behavior and to wear an appropriate gown. In 1263, he was in trouble again. He was absent from the register entries during that year, yet he had submitted a letter to the archbishop in which he admitted to having a sexual affair with one of his parishioners, in addition to excessive drinking. Like Gervaise, he promised to resign if he was found guilty of this in the future.[43] These cases strongly suggest that acquiring a benefice and being ordained a priest did not necessarily grant social adulthood. It would have been difficult for parishioners to view these particular priests with the honorable status of a married man (to the Church) and householder (of the parish) when they engaged in such disorderly behavior, behavior equated with youth. Daniel Bornstein has shown that in the rural parishes of Cortona, priests who cohabitated with women were still shown respect by their communities so long as they still adequately performed their job duties.[44] Thus, concubinage, while still outside of the institutional definition of clerical masculinity, was not seen as a problem by some parish communities which could accord honor and status to this relationship. But the behavior of Gervaise and William indicates a reluctance to be publicly identified with the priesthood, and it is difficult to imagine that they adequately performed their job duties. Furthermore, it must be recalled that the notations in Odo's register came from reports and interviews with the community. Clearly, their communities were not pleased with the behavior of their curates.

Is it at all possible that these clerics engaged in such errant behavior because they were unlearned? Could university training have modified the behavior of these clerics? This is a possibility for parish clerics, some of whom may have only had a rudimentary education in basic reading and writing. While a university education appeared to some contemporaries as the antidote to youthful, rambunctious behavior, providing a solid intellectual foundation in the pastoral life, one must also remember that university life in the Middle Ages provided extraordinary opportunities for transgressive behavior. Attending the university may not have helped mature the cleric, but instead bring about a contrary outcome. While one may have increased one's knowledge of their clerical vocation, there was ample opportunity to engage in disorderly behavior. Adam, priest at Yvecrique, came to his position sometime during the episcopate of Maurice (1231–5). In 1248, he was rumored of sexual promiscuity with two women and frequenting the taverns; he was also noted for his rebelliousness (*rebellis est decano*) toward the rural dean and non-residence in his parish. At this time, Adam might have been as young as 37 years of age. Three years later, he obtained a license from the archbishop to attend university, and subsequently earned a master's degree. The next mention of the cleric in Odo's register is in 1260, when we expect that Adam would have been somewhere in his fifties. Adam's rebellious nature was again noted. The priest had been excommunicated three times. He had not celebrated Mass for nearly a year, and the last occasion that he attended to his benefice was at Christmas time. Yet he had been in the tavern with friends the previous week. Although Adam was cited to appear before Odo for these and other charges, he did not. By the time Adam was finally deprived of his benefice in 1262, *in absentia*, he would have been over 50 years old.[45]

The example of Magister Robert of Houssaye, the rector at Conteville, further confirms this dynamic. As early as 1248, Robert was identified as a *magister*, indicating that he had attained a master's degree, and so had acquired an advanced university education; yet Robert was not a priest. The first time he was cited for his errant behavior was in 1248, when Odo's scribe noted the cleric was under suspicion for drunkenness, sexual misbehavior, non-residence and harassing the villagers. Aside from behaving badly, Robert refused to proceed to the order of priest. After being repeatedly warned, the archbishop ordered the cleric to resign. Robert refused and eventually was deprived of his benefice. At the time of his last citation in the register, 1263 (15 years later), Robert likely would have been between the ages of 44 and 52.[46] Looking at Robert's behavior, it seems quite obvious that clerical masculinity was not the most compelling gender performance for him. Yet, Robert did not want to resign his position. Why then did he refuse to be ordained a priest, which would only have solidified his benefice as a permanent position?

Perhaps Robert's case allows us the most remote glimpse of the medieval psyche, which the following case illuminates more fully.

It has been elsewhere noted that some clerics delayed ordination to the priesthood until a benefice was secured.[47] Odo's visitation records, however, demonstrate the opposite: beneficed clerics delaying ordination to the priesthood, as in the case of Robert of Houssaye. Canon law did not require that a rector be a priest, although in such a case a priest had to be employed by the rector to perform the sacramental duties. Clerics in minor orders sometimes obtained church benefices with the expectation that they would proceed in a timely fashion to the order of the priesthood. This appears to be the case with Roger of Sorenc. Under investigation in June 1253, Roger, the rector of the church at Limésy, agreed to relinquish his benefice if he did not proceed to Holy Orders. Eight months later, Odo's emissary ordered Roger to pay a fine in this matter or else face a formal proceeding. While there is no way to exactly pinpoint Roger's age, he was old enough to enter the priesthood, and the records do not indicate that a dispensation was made in his case. Roger momentarily disappeared from Odo's register, and resurfaced 7 years later, in 1260. Two noticeable items concerning Roger in 1260: first, he was noted as *presbyter*, indicating that he had finally been ordained a priest; and second, he appeared in Odo's register because he was under investigation for an adulterous affair with Mathilda, the wife of another cleric.[48] Did Roger put off becoming a priest because he was reluctant to make a lifelong commitment to celibacy? Was he trying to postpone the acquisition of spiritual authority, in effect social adulthood? It leads one to believe that perhaps Roger was cognizant of the enormous responsibilities that awaited him as priest, as well as the essential change in masculine status. The masculine self known before ordination would cease to exist in favor of the masculine self created after entry into the priesthood.

Why would these men accept a church benefice yet refuse to be ordained to the priesthood as required? Even in thirteenth-century Normandy, many elite families controlled the patronage of their parish churches, and thereby controlled the selection of those who would acquire that benefice. In order to keep the church under the control of the family, younger sons would acquire the benefice and become the rector of the church. These men were left out of the noble titles and vast family wealth their older brothers inherited; the promise of a church benefice, a comfortable life, would have satisfied most had it not been for the restrictions on their personal behavior. One might think that when a father passed wealth and property to his son, in this case, a parish benefice, that this would have signified a marker of adulthood. After all, for a priest, this livelihood came the closest to the inheritance of first-born sons. William, priest of the parish church at Nêle-Normandeuse, is a prime example of this scenario. William's father was the patron of this parish church, and from William's *bachelier*-like behavior, it becomes clear that he was descendent from an aristocratic family. William was running around with women and drinking excessively. His actions sometimes were violent, as he had engaged a knight in a duel and even assaulted his own father.[49] In short, William was behaving like a *bachelier*, a knight during his 'wild oats' period. Although he had acquired the social equivalent of a household in the form of a parish benefice, it seems as though he refused to grow up.

For some beneficed clerics, stability of position did not promote clerical manhood. Thus, as these cases demonstrated, the paradigm set out by Cullum did not always apply to medieval Norman clerics. Why did ordination to the priesthood not result in spiritual and social adulthood for these men? The answer to this question may lie in the clerics' perception that the clerical model was deficient. Like the unbeneficed clergy, the bad behavior of beneficed priests was reflective of the ways in which clerics tried to negotiate their self-identities according to the external demands of their peers. These clerics must have perceived deficiencies in the model of clerical manhood that they were to assume. Thus, they acted like youths because they did not perceive spiritual marriage and spiritual fatherhood to be true, effective markers of masculine adulthood. For these men, biological fatherhood and marriage was likely perceived as a legitimate, real marker of masculine adulthood. They continued to behave like adolescents because for them the honor and status of medieval adulthood was not really acquired.

IV Becoming a man

Why did some priests make their passage into social and spiritual adulthood, accepting new clerical identities while others floundered in adolescent behavior? Why was it permanent for some, but not for others? There is likely more than one reason for the adolescent behavior of these men of the Church. Social origins and environment must have played some role. At the most basic level, however, the failed negotiation of a new gender identity provides the most compelling reason why these men failed to adjust to their proper clerical roles. It must be remembered that clerical masculinity was an artificial construct, one created by an institution for imposition on a select group of society. There was an immediate disjuncture between the ideals of clerical manhood and secular notions of masculinity. Men in the secular world lived their lives according to the ideals of their social class and life-cycle. Values of their social status, like fornication and shedding blood for the nobility, did not have a legitimate role in the clerical life-cycle.

The assumption of spiritual authority and concomitant clerical adulthood was predicated on a priest willingly embracing the ideals of clerical masculinity. The argument for an extended adolescence makes better sense if we assume that ordination to the priesthood and the acquisition of a parish benefice did not automatically end adolescence for these men nor did it immediately confer adulthood. Rather, for some parish clerics in the diocese of Rouen, social adulthood was delayed as these clerics negotiated their gender identities. To be received as an adult, one's status must have been stable; yet one must also be received by the community as socially adult. Priests, even beneficed ones, who behaved as youths by taking part in immature behavior, would not have been received as adults by the parishioners in their communities; the frequency with which clerics were reported to the archbishop indicates community disapproval. Social adulthood depended upon the community granting adult status. It would be expected that after a transitional phase, some 'youthful' priests might have eventually embraced the duties and responsibilities of their vocation. Yet, in many cases indicated by Odo Rigaldus' register, the transition into clerical manhood was not successful. Instead of acquiring adult male status by the acceptance of this new gender identity, some clerics who entered the priesthood simply continued to behave in accordance with peer approval. They denied spiritual 'fatherhood' and the higher status of a curate. The cases from Odo's register indicate that some parish clerics did not make adjustments to clerical manhood by creating spiritual families and taking on social adulthood. These men were either unwilling or unable to acquire clerical manhood and adult status. They instead negotiated their clerical identities within a context of masculine bonding.

To truly understand gender identity and the clergy, we as scholars should remember that this was not a monolithic group. The clerical life-cycle would have been varied between secular and monastic, beneficed and unbeneficed and minor and major cleric. Clerical manhood as an ideal existed, but its manifestation was clearly variable. It is also important to remember that as it was promoted, it was in constant tension with secular versions of manhood, masculinities that ordinary men lived by and that still had compelling significance for clerics.

Notes

1. Bibiliotheque Nationale, MS Latin 1245; *Regestrum visitationum archiepiscopi Rothomagensis*, edited by Theodosius Bonnin (Rouen, 1852); an English edition, currently out of print, is *The Register of Eudes of Rouen*, translated by Sydney M. Brown and edited by Jeremiah F. O'Sullivan (New York, 1964). All references are to the Latin edition, abbreviated Bonnin.

2. Bonnin, 116, 187, 285, 409, 411, 583, 652. My conclusion is based on the number of years that both men appeared in Odo's register. For Ferric, who was also the parish priest of Saint-Pierre Vieux de Dun, age can be roughly calculated by the minimum age requirement for the priesthood, which was 25 years, except in cases of papal dispensation. This would have made Ferric approximately 41 years old by the time of his last appearance in the register.

3. P.H. Cullum, 'Life-Cycle and Life-Course in a Clerical and Celibate Milieu: Northern England in the Later Middle Ages', in *Time and Eternity: The Medieval Discourse*, ed. Gerhard Jaritz and Gerson Moreno-Riano (Turnhout, 2003), 271–81.

4. See P.H. Cullum, 'Clergy, Masculinity and Transgression in Late Medieval England', in *Masculinity in Medieval Europe*, ed. D.M. Hadley (New York, 1999), 178–96; in the same collection, R.N. Swanson, 'Angels Incarnate: Clergy and Masculinity from Gregorian Reform to Reformation', 160–77; and Jennifer D. Thibodeaux, 'Man of the Church or Man of the Village?: Gender and the Parish Clergy in Medieval Normandy', *Gender and History* 18/2 (August 2006): 380–99. See also the essays in this volume by Anthony Perron and Janelle Werner.

5. The *Polyptychum Diocesis Rothomagensis* was begun by one of Odo's predecessors, Archbishop Pierre Colmieu (1236–44); both Odo and his successor Guillaume de Flavacourt added information to it. The census lists every parish in the diocese of Rouen, providing information on the name of the rector, the patron of the church, the number of households in the parish and the value of the benefice. Thus, one can learn the names of the rectors instituted to the parish benefice and the names of the archbishops who instituted them. The census was not always updated in every parish and in some cases the names of the rectors were not listed. Unlike the Odo's register, which contains dated entries, the census does not provide us with the dates when it was last updated. See *Polyptychum Diocesis Rothomagensis*, in *Recueil des historiens des Gaules et de la France*, volume 23, ed. De Wailley et al. (Paris, 1874).

6. For more on this classification, see Michel Pastoureau, 'Emblems of Youth: Young People in Medieval Imagery', in *A History of Young People in the West, volume I: Ancient and Medieval Rites of Passage*, ed. Giovanni Levi and Jean-Claude Schmitt (Cambridge, 1997), 223.

7. Levi and Schmitt, *A History of Young People in the West*, 7–8.

8. Barbara A. Hanawalt, *Growing Up in Medieval London: The Experience of Childhood in History* (Oxford, 1993), 10–13. Hanawalt is one of the few historians to address the female childhood and adolescent experience.

9. Georges Duby, 'In Northwestern France: The Youth in Twelfth-Century Aristocratic Society', in *Lordship and Community in Medieval* Europe, ed. Fredric

L. Cheyette (New York, 1968), 199. William Marshal's extended adolescence ended in 1189, when he married at the age of 45.

10. Another well-documented group was the artisans, considered to be socially young as they lived in the homes of their masters, unable to marry and acquire their own household until they completed their apprenticeship. Such training could begin as early as age 10 and as late as age 21. The apprentice was in a state of dependence, unable to exercise adulthood while in training. Social adulthood was achieved for these men with marriage, the acquisition of their own household and the acquisition of their own workshop. See Ruth Mazo Karras, *From Boys to Men: Formations of Masculinity in Late Medieval Europe* (Philadelphia, 2003), chapter 4; P.J.P. Goldberg, 'Masters and Men in Later Medieval England', in *Masculinity in Medieval Europe*, 56–70.

11. Deborah Youngs, *The Life Cycle in Western Europe, c. 1300–c. 1500* (Manchester, 2006), 131–2, calls marriage the 'closest to a universal rite of passage' to adulthood.

12. For some examples, see Shannon McSheffrey, 'Men and Masculinity in Late Medieval London Civic Culture: Governance, Patriarchy and Reputation', in *Conflicting Identities and Multiple Masculinities: Men in the Medieval West*, ed. Jacqueline Murray (New York, 1999), 243–78; and Susan Mosher Stuard, 'Burdens of Matrimony: Husbanding and Gender in Medieval Italy', in *Medieval Masculinities: Regarding Men in the Middle Ages*, ed. Clare A. Lees (Minneapolis, 1994), 61–72.

13. Youngs, 119.

14. Georges Duby, 'Youth'; Ruth Karras, *From Boys to Men*, chapters 3 and 4; and Jacques Rossiaud, *Medieval Prostitution* (Oxford, 1988).

15. Ordericus Vitalis, *The Ecclesiastical History of England and Normandy*, translated by Thomas Forester (New York, 1968), volume 2, 171; see also W.M. Aird, 'Frustrated Masculinity: The Relationship between William the Conqueror and his Eldest Son', in *Masculinity in Medieval Europe*, 39–55.

16. Heinrich Denifle, *Chartularium Universitatis Parisiensis*, volume I (Paris,1889), 481–2.

17. Chartularium, I: 540–1.

18. Hanawalt, *Growing Up in Medieval London*, 196.

19. See Jo Ann McNamara, 'The "*Herrenfrage*": The Restructuring of the Gender System, 1050–1150', in *Medieval Masculinities*, 3–30; and Swanson, 'Angels Incarnate'.

20. See note 3.

21. Patricia Cullum first addressed the issue of a different 'life-course' for the clergy in 'Life-Cycle and Life-Course in a Clerical and Celibate Milieu: Northern England in the Later Middle Ages', in *Time and Eternity*, 272–81; this analysis was further expanded in 'Boy/Man into Clerk/Priest: The Making of the Late Medieval Clergy', in *Rites of Passage: Cultures of Transition in the Fourteenth Century*, ed. Nicola McDonald and W.M. Ormrod (York, 2004), 51–65.

22. John Shinners and and William J. Dohar, eds., *Pastors and the Care of Souls in Medieval England*, 50; Cullum, 'Boy/Man into Clerk/Priest', 55.

23. Dyan Elliott, *Fallen Bodies: Pollution, Sexuality and Demonology in the Middle Ages* (Philadelphia, 1999), chapter 4.

24. Shinners and Dohar, 50–4. The minimum age for a candidate to the priesthood was 23, but this was only possible with a papal dispensation. Canon law specified 25 years as the minimum age for the ordination to the priesthood.
25. For a summary of these changes, see Shinners and Dohar, 121–2.
26. P.H. Cullum, 'Life-Cycle and Life Course in a Clerical and Celibate Milieu: Northern England in the Later Middle Ages', in *Time and Eternity*, 274–81; also 'Clergy, Masculinity, Transgression', in *Masculinity in Medieval Europe*, 192–6.
27. See the essay in this volume by Derek Neal.
28. Cullum, 'Boy/Man into Clerk/Priest', 58.
29. Cullum, 'Life-Cycle and Life Course', 276–8; this idea, however, is more likely accurate for late medieval towns, than thirteenth-century rural Norman parishes. Priests in small Norman villages were less likely to establish a spiritual *familia* within their households, because of the small size of their parish church, land and home. Undoubtedly some of the wealthier parishes did have a full staff in charge of the church, but the census for the diocese of Rouen indicates considerable variance in the size and wealth of parish churches.
30. Youngs, *The Life-Cycle in Western Europe*, 132, 152.
31. Lateran IV, canon 32.
32. For example, see John W. Baldwin, *Masters, Princes and Merchants: The Social Views of Peter the Chanter and His Circle*, volume one (Princeton, 1970), 91, on the participation of subdeacons in the Feast of Fools.
33. Bonnin, 466.
34. Bonnin, 549.
35. Bonnin, 587.
36. Cullum, 'Life-Cycle and Life-Course in a Clerical and Celibate Milieu: Northern England in the Later Middle Ages', 279.
37. Karras, *From Boys to Men*, 63.
38. Bonnin, 62, 297.
39. To confirm this hypothesis, the ages of a select group of parish priests were calculated using the data supplied by the census for the diocese of Rouen, and the information located in Odo's register. Taking into consideration the minimum age requirements for ordination to the priesthood, I was able to approximately pinpoint the age range for the case studies in this analysis. My cases cited here indicate that the cleric under analysis was a priest (not a deacon or subdeacon or cleric in minor orders), and since 25 was the minimum age for entry into the priesthood, I was able to use this framework for calculating age. There exists the possibility that the cleric in question had received papal dispensation to bypass the mandatory age. For instance, the pope granted Nicholas de Hermeville such a dispensation in 1253, although Nicholas was only 20 years old (Bonnin, 176). In order to further demonstrate my methodology, it is necessary to first provide an overview of the process of institution to a benefice. There was a specific procedure involved with priests acquiring benefices. The census provided me with the archbishop that instituted the priest in question. First, priests were selected by the lay or ecclesiastical patron who held the advowson to the parish church. Once selected, they were presented to the archbishop (or his proxy) for

formal institution to the benefice. In the census for the diocese of Rouen, six archbishops were recorded for their institution of priests to benefices. Taking into consideration the dates of the archbishop's reign, and the minimum age for ordination to the priesthood, I was able to then project an age range for the priest in question at the time of his citation in Odo's register.

40. Bonnin, 116, 145, 187, 209, 285, 409, 411, 583, 652.
41. For William and Ralph of the church at Bailly-en-Rivière, see Bonnin, 27 and *Polyptychum*, 263; For Matthew and Baldwin of the church at Saint-Rémy-en-Campagne, see Bonnin, 330, 475, 479, 610, 614, 660, 661 and in *Polyptychum*, 261–2.
42. *Polyptychum*, 265; Bonnin, 19–21. The *Polyptychum* also lists a Magister Nicholas instituted to the same church. Ordination lists from the register list Magister Nicholas, 'persona Sancti Remigii in Riparia', ordained to deacon and then priest in 1263 (Bonnin, 687). This would indicate that Gervaise has either resigned or died.
43. Bonnin, 27, 328–9, 668; also *Polyptychum*, 264.
44. Daniel E. Bornstein, 'Parish Priests in Late Medieval Cortona: The Urban and Rural Clergy', *Quaderni di Storia Religiosa* 4 (1997): 165–93.
45. Bonnin, 30, 119, 385, 434, 660; *Polyptychum*, 296.
46. Bonnin, 22–3, 406–7, 475, 479, 655, 656.
47. Cullum, 'Life-Cycle', 274; R.N. Swanson, *Church and Society in Late Medieval England* (Oxford, 1989), 42–3.
48. Bonnin, 160, 205, 385–6, 653, 663. The seriousness of the charge required Roger to undergo canonical purgation with the tenth hand of priests.
49. Bonnin, 19, 22.

7
Promiscuous Priests and Vicarage Children: Clerical Sexuality and Masculinity in Late Medieval England

Janelle Werner

In 1501, a vicar named Roger Homme was summoned to appear before the church court of the diocese of Hereford. The judge charged him with sexual incontinence, alleging that he had had sex with Isabel Herford, a married woman from a neighboring village. Despite Roger's vow of chastity and Isabel's marital status, theirs was no casual relationship. According to the charge, the two lovers behaved as husband and wife: 'Sir Roger Homme, vicar of Canon Frome, is incontinent with Isabel Herford, wife of James Herford of Munsley, whom he holds [and] from whom he produced a child, as if he has married her.'[1]

Roger Homme and other priests who had either brief affairs or longer, marital-like relationships with women expose a neglected issue in the history of the late medieval clergy. Historians, assuming that priests complied with the ecclesiastical mandate of celibacy, have downplayed clerical sexuality and its ramifications for clerical masculinity. Because the standard markers of lay masculinity – sexual virility, skill in combat and heading a household – were unavailable to them, celibate priests, the argument goes, carved out new ways of being masculine. Unchaste priests were probably in the minority, but they constituted a visible element of the priesthood, especially among the parish clergy. After all, who in the small, remote village of Canon Frome could have been unaware of Roger's relationship with Isabel? It may seem like common sense that *all* priests were not *always* celibate, but this overlooked aspect of clerical life has implications for conceptions of clerical identity and masculinity.

Historians continue to debate the full extent to which priests had relationships with women, but unchaste clergy on the European Continent have been more forthrightly acknowledged and studied than those in England. Most scholars agree that fornication and concubinage were relatively common, perhaps particularly in the Low Countries, France, Germany, Spain and Italy.[2] English thinkers, however, have long looked across the Channel and perceived themselves as different. Defending English clerics against criticism, Thomas More asserted:

> But yet for that I have myself seen and by credible folk have heard, like as you say by our temporality, that we be as good and as honest as anywhere else…the secular clergy is in learning and honest living well able to match and…far able to over-match number for number the spirituality of any Christian nation.[3]

Today, most English historians would still agree, arguing (or assuming) that English priests were better behaved than their continental peers.

Current research on clerical concubinage in late medieval England stands in stark contrast to the wealth of quantitative and qualitative studies of priests, their concubines and their communities on the Continent, and the most frequently cited studies of clerical incontinence are based on limited samples of the documentary evidence. In his study of sixteenth-century ecclesiastical courts, Ralph Houlbrooke looked at accusations of sexual misconduct against clerics in two archdeaconries in the dioceses of Winchester and Norwich, but only examined records from one year in each diocese. He found 11 clerics charged with incontinence in the diocese of Winchester between 1527 and 1528, and eight clerics in Norwich in 1538. While Houlbrooke admitted that his records were incomplete, he nonetheless broadly concluded that incontinence among the clergy was rare. Tim Cooper, in his monograph on the pre-Reformation clergy, cast a slightly larger net in the records of the Lichfield consistory court: between 1524 and 1531, charges were brought against 18 clerics in the diocese. While he acknowledged that these charges might 'confirm negative stereotypes of the parish clergy', he nonetheless argued, like Houlbrooke, that 'the problem [of incontinence] was comparatively small-scale'. Margaret Bowker, in her study of the diocese of Lincoln, surveyed clerical incontinence in four different years over the course of the early sixteenth century. Sampling four Lincoln court books from the early sixteenth century, she found varying statistics, with more charges toward the end of her survey period: between 1514 and 1521, priests in 10 per cent of parishes visited were charged with incontinence. Because good comparative figures are not available for the sixteenth century, Bowker rightly refused to make broad generalizations about the extent of sexual immorality among the clergy, but other historians, such as Christopher Harper-Bill and Peter Marshall, have not hesitated to use her statistics to do just that.[4]

Peter Heath's study of the diocese of York uniquely examined clerical incontinence over a comparatively long period of time, but his geographical area was still limited, covering only between 50 and 86 parishes (the number varies by date) within the diocese. Heath examined one late fifteenth-century court book (1453 to 1491) from the Dean and Chapter of York and found 93 charges against priests for fornication or adultery over a period of 40 years. He concluded, based on dropped charges and purgation, that only 1.5 offenders per year were guilty of sexual misconduct and that, therefore, sexual offenses were uncommon and easily corrected by the church courts.[5] Even Heath's statistical analysis, while more inclusive than the rest of the English studies, was based on a relatively narrow sample of the available evidence. None of these brief studies considered variations over time, for example, or whether particularly zealous or lax prosecution of clerical incontinence might affect reported numbers.[6]

This narrative of English 'exceptionalism' has been used to frame topics as diverse as the English feudal system, the development of common law and the history of Parliament, but it has had a particularly tight grasp on the history of religion. A far-reaching challenge to this orthodoxy has recently been presented by Kathyrn Kerby-Fulton in *Books Under Suspicion*, in which she argued that English religious thought had a long history of intellectual radicalism that was not simply based on Wyclif and his followers, but was, instead, heavily influenced by continental writers. Kerby-Fulton queried the usefulness of the pervasive England/Continent binary, concluding that English religious culture was more pluralist and less insular than scholars have assumed.[7] I extend Kerby-Fulton's challenge of English exceptionalism to another aspect of religious culture and argue that priests who committed fornication or kept concubines were as common in England as on the Continent.

Even while continental scholars have examined clerical sexual misbehavior, few have discussed its implications for clerical masculinity; most scholarship on clerical masculinity has focused on the ways by which clerics were *distinguished* from laymen. Robert Mills argued that tonsure acted as a visible signifier not only of clerical status (differentiating clerics from laymen), but also of religious authority. Maureen Miller maintained that after the ban on clerical marriage, the clergy were denied an 'outward marker' of lay masculinity. Rather than negating their masculinity, clerics' distance from women let them attempt to define 'an extreme masculinity', one more powerful than lay masculinity because 'it was not weakened by association with the weaker sex'. According to Jacqueline Murray, clerical masculinity and secular masculinity were contradictory belief systems.[8] Secular masculine identity was defined in terms of military prowess and sexual virility, while clerical masculinity 'eschewed both'. Celibate men, Murray claimed, attained masculinity by redefining it: clerics (monks, in particular) appropriated military language, framing the achievement of chastity as a battle against lustful temptations and redirecting military values to their spiritual struggles.[9]

Assuming that clerics embraced and practiced celibacy, English scholarship, in particular, has left little room for an exploration of ways in which characteristics of clerical and secular masculinities might have overlapped, especially in terms of sexuality. Like continental scholars, English historians have argued that clerical masculinity was distinct from lay masculinity. Robert Swanson and Patricia Cullum both highlighted the incompatibility of lay and clerical models of masculinity. As Swanson put it, clerics were *male*, but they were not *men*; they were, instead, 'emasculine'. Having renounced masculinity, clerics achieved a 'genderless status'. Because they could not marry, have children or carry weapons, parish clergy, in particular, were caught between norms of lay masculinity and their status as a third, emasculine gender. Their chastity challenged normal social patterns, 'particularly threatening familial and household relationships between men and women as rulers and subjects'. Because they were denied marriage, clerics could also not fully participate in social relationships. Cullum identified a fundamental conflict between lay and clerical masculinities, arguing that young clergy were forced to make a choice: 'to keep their vows and risk their masculinity; or to confirm their masculinity at the expense of their vows'. Cullum further argued that, not only were clerics not fully masculine, they were not fully socially adult, either, because they did not marry and become heads of independent households.[10]

Instead of taking celibacy as a given, I propose to explore the possibility that clerical masculine identity was not always distinct from – or at odds with – lay masculinity. Based on evidence from consistory court and episcopal visitation records of the late medieval diocese of Hereford, this essay argues that secular clerics had sexual relationships with women more frequently than scholars have recognized and that in these relationships, priests often behaved like laymen and adopted elements of lay masculinity. For a few priests, celibacy was not even a component of their *clerical* identity.

I The diocese of Hereford and its records

Medieval England had no such thing as a typical diocese. Some bishoprics were secular sees, while others were monastic, staffed by Benedictine monks or Augustinian canons. The dioceses of York, Lincoln, Norwich and Coventry and Lichfield were geographically vast, each encompassing several counties; Ely and Carlisle were minute in comparison. The diocese of London was small but densely populated; in contrast, the population of the diocese of York was spread thin, except for the town of York itself. Some dioceses were wealthy, but a priest in the diocese of Coventry and Lichfield most likely held a benefice worth less than £10.[11]

This essay takes the diocese of Hereford as a case study. Located in the English West Midlands, the medieval diocese was situated next to – and partially straddled – the Welsh–English border. Although Hereford is no more *typical* than any other diocese, its size, population, wealth and location were *characteristic* of many English dioceses. Hereford's especially rich sources, including a long series of court books and a set of visitation returns, make it an excellent locale for an exploration of parish life in late medieval England.

The diocese of Hereford was roughly contiguous with the county of Herefordshire, but also incorporated parishes from Wales and other counties in the Marches.[12] It was a mid-sized diocese, much smaller than the largest English dioceses, but roughly on par with Worcester, Winchester and others.[13] Difficult terrain and border warfare meant that the Marches were sparsely populated throughout the medieval and early modern periods. In this rolling landscape, most people (four of every five) lived in small, nucleated villages, and a handful lived in even more remote settlements. The town of Hereford was the largest in the area, with a 1377 population of 3000–4000, roughly comparable to Worcester, Cambridge or Leicester. Hereford's population pattern was typical of the West Midlands, which tended to have more numerous small towns than other English counties.[14]

According to the *Valor Ecclesiasticus* – an assessment of benefice values conducted in 1535 for the purpose of taxation – Hereford's clergy were not particularly affluent. Hereford ranked 15 out of 16 shires or dioceses in terms of the average values of rectories, and 13 in terms of the average values of vicarages. For example, the average value of a benefice in Essex, in the top third of these rankings, was about £14, while Hereford's average was only £8. Although these rankings suggest that Hereford clergy were exceptionally poorly provided for, many English clerics were not well off. In the neighboring diocese of Coventry and Lichfield, 77 per cent of vicarages and almost 50 per cent of rectories were worth less than £10.[15]

The Marches was an area of cultural contact between the Welsh and the English, and Hereford's location near the border distinguished it from many English dioceses.[16] Yet, the melding of different populations occurred not only in border territories like the Marches but also in other areas of high migration and cultural assimilation, like London and Bristol. Hereford's position in the Marches meant that border migration and warfare were enduring aspects of parish life, as they were in other areas that were near Wales (Coventry and Lichfield, Bath and Wells, and Worcester), and perhaps even the dioceses of Durham and Carlisle, located in the Scottish Borders.[17]

Certain characteristics of Hereford were singular, but it was nonetheless a diocese that was roughly average in terms of its size, population density and wealth and was certainly more characteristic of English dioceses generally than the densely populated and wealthy dioceses of east and southeast England. A few well-documented dioceses – Lincoln, York and London – have been studied extensively and are often taken as representative English dioceses. But as Robert Swanson has noted, a diocese with small towns, large parishes and an agricultural economy, such as Hereford, was 'typical of an important group among English parishes, but perhaps not typical of those for which most information survives'.[18] The diocese of Hereford, although on the geographical fringe of England, was hardly peripheral.

Evidence of clerical sexual misbehavior in the diocese of Hereford appears primarily in the returns of an episcopal visitation and act books from the consistory courts. Both the visitation, conducted in 1397, and the fifteenth-century act books recorded bishops' efforts to prosecute unchaste clerics and their sexual partners. Neglected until fairly recently, ecclesiastical records are now recognized as an untapped and useful source for evidence about medieval marriage and families. Scholars have scoured act books and depositions for evidence about courtship, marriage, separation and divorce, but these records are just as informative about irregular marriages and marital-like unions, including clerical relationships.[19] Ecclesiastical records, however, are not unproblematic. Both visitation returns and act books tended to be terse and highly formulaic; frustratingly, they often did not record the outcome of a case.

Yet they are indispensible, too, providing a rare glimpse into the lives of the parish clergy. As with all medieval records, there is always uncertainty: we cannot know how many transgressions slipped through the net of the visitation or went unseen by church court officers. We cannot even know all the accusations were valid. What we do know, however, is what was reported.

II Clerical fornication and concubinage

The best data about clerical sexual misbehavior in Hereford comes from its fourteenth-century episcopal visitation. In 1397, Bishop Trefnant and his representatives traveled through the diocese, parish by parish, questioning local clergy and lay juries about the maintenance of their parish churches and the moral lapses of their peers. Parishioners complained about adulterous neighbors, neglected churches and missing service books; they also commented on the behavior of their parish clergy.[20]

Their complaints, coupled with similar offenses noted in fifteenth-century court records, suggest that the English clergy broke their vows of celibacy as often as priests on the Continent. As with all estimates of medieval populations, we can only approximate the number of clerics in the late medieval diocese of Hereford, but there were roughly 1000 secular priests in the entire diocese and 470 priests in the portion of the diocese covered by the 1397 visitation.[21] In total, 67 secular clerics, approximately 14 per cent of the clerical population, were named in cases involving sexual misconduct during Trefnant's visitation.[22] The remaining ten clerics named in the visitation were regular clergy (that is, men who lived within monastic houses), totaling 8 per cent of the population of regular clergy.

This is one way to look at the issue; another is to consider how many parishes had a misbehaving priest in their midst. Out of 260 parishes and chapels reporting in the visitation, 55 (or one out of every five parishes) had one or more unchaste priests. Of those, 19 parishes – one out of every 14 parishes – reported priests who held concubines. A further nine parishes were homes to women who were accused of having sex with priests who lived in other villages or dioceses. All told, every fourth parish had an unchaste priest, a priest who held a concubine, or a woman accused of sleeping with a cleric; other clerical couples may have gone unreported. And even those villagers who did not have a misbehaving priest in their midst would probably have known about such relationships, either by coming in contact with a priest and his lover in a neighboring village or through networks of trade or gossip.

Not only were more parish priests charged with sexual incontinence than monks, but regular and secular clerics also had distinct patterns of sexual misconduct. Monks were charged almost exclusively with fornication, with only one man accused of concubinage (if guilty, he was also unfaithful, for he was additionally accused of fornication with another woman). Seventy per cent of all secular priests named in cases of sexual misbehavior were charged with fornication, while 27 per cent were charged with concubinage, accused of holding a 'concubine' or 'keeping a woman'. Secular priests predominantly offended through fornication, but they were still significantly more likely than monks to keep a concubine.

Secular clerics also behaved more like laymen than like monks. Of the 463 charges of sexual misconduct against laymen in the visitation, 32 per cent targeted on-going relationships. Likewise, 28 per cent of charges against secular clerics involved concubinage. Parish priests and chaplains, then, behaved differently from monks, but quite similarly to the laymen who were their parishioners and neighbors.[23]

English historians have traditionally downplayed clerical fornication and concubinage. Yet statistics from Hereford's extensive records suggest that clerical sexual misconduct in England was comparable to that on the Continent. A study of records from the diocese of Tournai has shown that between 6 per cent and 11 per cent of curates were fined for incontinence in the late medieval period; between 12 per cent and 34 per cent of the secular clergy in the thirteenth-century diocese of Rouen were accused of errant sexual behavior, depending on the geographic area.[24] While priests who misbehaved sexually were not a majority of the English clergy, they nonetheless made up a significant portion of the clerical population.[25]

Because an episcopal visitation was a systematic and, ideally, exhaustive examination of the diocese, Hereford's visitation returns provide a thorough picture of clerical incontinence at one moment in time. After 1397, we must rely on the records of the consistory court. Each diocese in England held a consistory court, the proceedings of which were recorded in an act book. As ecclesiastical courts, they held jurisdiction over the moral infractions committed by clergy and laity, such as non-attendance at church services, defamation, usury, sorcery and, especially, sexual misbehavior of both laity and clergy.[26] These records, like the visitation, show that priestly incontinence was still common throughout the fifteenth century, though perhaps less so. This qualification, however, might be more an artifact of the sources than a reflection of practice. Unlike episcopal visitations, church courts were held year-round, but charges were made only when churchwardens, prominent parishioners or local clergy cared enough to bring an offense to the court's attention; as a result, act books were not as likely to catch as many offenders as the bishop's representatives did when they visited each parish.[27]

Statistics from consistory court books from four sample years show that charges of clerical sexual misbehavior were less common in the 1400s than in 1397, but steadily increased over the course of the fifteenth century. In the court year spanning 1445–46, 16 charges of fornication and six of concubinage were brought against clerics. In 1468–69, more priests were accused of having on-going relationships with women: while there were only 14 accusations of fornication, there were 19 cases of concubinage. These figures continued to increase in the late fifteenth century. In 1487–88, priests faced 32 accusations of fornication and 20 accusations of concubinage; by 1501–02, these numbers had increased by almost half, with 53 charges of fornication and 28 of concubinage. Perhaps more priests were committing indiscretions, or perhaps court officers were increasingly interested in prosecuting these cases. Even though their misbehavior was reported to the consistory court in a somewhat haphazard fashion, a steadily growing number of priests was brought before the court each year.[28]

III Clercial sexuality and masculinity

Consistory court books not only show that prosecution of incontinent priests increased during the course of the fifteenth century, they also tell us about clerical sexual relationships. Priests who came to the attention of the court were charged with a variety of sexual misdeeds. Like laymen, priests had brief sexual affairs: John Dyer, a chaplain, fornicated with a woman named Lusot in 1468. Accused of having sex with Katherine Onislow, Joan Bennett, Alice Coke and Maud Baghe from 1501–02, vicar John Ball of King's Pyon seems to have been something of a serial monogamist. Like laymen, priests committed adultery: in the same year, a chaplain named Maurice committed adultery with Maud Burton, a married woman. Like laymen, priests got women pregnant: in 1517, Anna Decons, a single woman, was summoned to court because she was pregnant and confessed that Christopher Wake, a chaplain in her parish, was the father of her child. And, like laymen, priests often ignored the warnings of court officers and continued committing their sin *in recidivo*, meaning (much like the word 'recidivist') that they had previously been charged, had perhaps been corrected, but nevertheless persisted in their error. In 1454, for example, John Baker, a priest, was charged with incontinence *in recidivo* with a woman named Isabel Gentill.[29]

Sometimes, priests were discreet (or at least tried to be discreet), like Roger Parler, who kept Joan Kiste 'in his room and house constantly and secretly' (*in camera et domo sua continue et secrete*). Other times, they were less prudent, like a vicar named Nicholas who, in 1468, was charged with incontinence with his concubine Cecilia, whom he held openly in his house (*quam tenet publice…in domo sua*).[30] Based on the frequency and variety of their sexual misconduct, many priests were indistinguishable from laymen in the church court records.

The range of sexual misdeeds committed by priests, then, was quite similar to that of their male parishioners. These similarities were not limited to sexual behavior alone, for priests also behaved *socially* like laymen; a significant minority of priests established stable, long-term relationships with women, relationships which very much resembled marriage. Some of these relationships lasted only a few years, while others persisted for decades. Thomas Nasshe, the rector of Munsley, and his concubine Joan Mawndfeld were brought into court in 1445 and again in 1454. In 1481, Lewis Hoptkyn was charged with holding a woman named Dyddge for more than 12 years. Owen ap Griffith, the vicar of the village of Meole Brace in Shropshire, lived with his concubine Alice Gridilston in the parish vicarage and had ten children with her by the time he was summoned to court in 1491. Owen and Alice must have been in a relationship for 20 years or more by the time they were brought into court.[31]

Like Owen ap Griffith and Alice Gridilston, clerical couples often lived together as if they were married. In 1473, a chaplain named Richard was charged with incontinence with his concubine, Alice, 'whom he held with himself in his house' (*quam tenet secum in domo sua*). Thomas Wilmotts and Maud Jeynkyns lived together in the same house, as did the curate of Shrawardine and his concubine. Although a chaplain named Thomas Gogh had been warned in February 1482 to evict his concubine, Margaret Bygolt, he confessed two months later in April, and then again in July, that they were still living together.[32]

Like laymen, priests fathered and raised children. John Davies had two children with his concubine, as did William Church and his concubine, Margaret. Priests' infants, like the children of laypeople, were sometimes sent to wet-nurses. Henry Mitchel, the vicar of Mansell Lacy, and Agnes Hopkyn had a child who had been sent to nurse with Margaret Catur, one of Mitchel's parishioners. John Carpynter, a vicar who had a child with a woman named Margaret, sent the child to live in the house of Thomas Blake in Bromyard, a town some twenty miles away. When they began to be kept in the 1530s, parish registers sometimes recorded the baptisms of clerical children alongside the children of laypeople: Ralph, the son of Elizabeth Patten and Hugh Holder, a priest, was baptized in the parish church of Bromyard in 1545.[33]

Living with women and raising children with them, these priests behaved much like husbands and householders. Even the language court officers used to describe these couples suggests that clerical relationships were understood to be similar to marriages. The records from Hereford occasionally used the term 'concubine' (*concubina*) to designate a priest's partner. More often, though, the court clerk appended the phrase *quam tenet* (or, 'whom he holds') to a charge in order to denote a long-term relationship. Use of the phrase *quam tenet* not only signified a lasting relationship, but also carried connotations of marriage, house-holding and economic power.

The phrase *quam tenet* spoke to the economic power of men (including priests), as well as to their sexual and social behaviors, for the verb *tenere* also had clear overtones of being a householder, a tenant – the economically dominant and socially responsible head of a domestic unit. In some cases, economic support extended to maintenance away from the household, for laymen were charged in Hereford with maintaining women in other houses, an arrangement that seems to have been especially common if he were already married. John Tewe, a married man who lived in Ludlow, committed adultery with Dylly Dege, whom he held at New Radnor, over 25 miles away. Priests did much the same, albeit to avoid scandal of a different kind. John Verne and his concubine Katherine did not live together, but their relationship was well-known to their parishioners (not least because they had sex in the parish church). John Hore, the vicar of Canon Frome, had a child with his concubine, Margery Tirrolde, 'whom he held in the house of William Tirrolde', probably Margery's father or brother.[34]

Concubines of priests often fulfilled the economic functions of a wife. A number of historians have pointed out that priests were frequently accused of having sexual relations with their female servants, and synodal legislation repeatedly warned priests not to live with any women, including servants, for just this reason.[35] John Glover, the vicar of Meole Brace, was deprived of his benefice in 1475 on account of a sexual relationship with his servant.[36] John Devereux made explicit the economic necessity of his concubine when, in 1477, his dean, Richard Pede, accused him of adultery with Margery Wynch, a married woman. John confessed to the crime of adultery with Margery, but when the dean asked him to renounce Margery and evict her from 'the house of their cohabitation', John hesitated. He pleaded that 'he could not renounce her at this moment…without serious harm because she is very necessary to him…in the service of governance of his house and household.' He vowed, however, to send her away 'immediately after the feast of St. Michael'. Margery was crucial, both economically and socially, to the proper functioning of John's clerical household.[37]

In terms of their sexual and social behavior, then, many priests adopted elements of secular masculinity. In fact, what most distinguished priests from laymen in the church court records was not their sexual behavior, but their punishment. Priests were punished less often and less harshly than laymen, and they nearly always avoided the usual penalty for sexual misbehavior – a public flogging. Most clerics were assigned a private penance or allowed to commute their penalty to a monetary fine.

Take, for example, the case of Thomas, the curate of Marstow. In 1469, Thomas confessed that he had fornicated with a woman named Joan. For penance, he was assigned not the usual punishment for a layperson – being flogged during a procession around the parish church – but a more private penance. Thomas was required to say seven penitential psalms, fast on bread and water for six weeks and make an offering of a candle at the shrine of St. Thomas in Hereford Cathedral. This was an extensive and long-lasting punishment, but did not include the ritual public humiliation of a public flogging and was performed almost entirely in private.[38]

Thomas' punishment was not unusual. Priests were punished less often for their misbehavior, and when they were punished, they received the chance to perform their penance privately far more often than laymen. In a sample of four deaneries during the 1468–69 court session, 181 laymen and 18 priests were charged with the sexual crimes of fornication and adultery (for laymen) or incontinence (for clerics). Of these laymen, 80 (44 per cent) either admitted their sin or were pronounced guilty by the judge and were assigned the penance of being publicly flogged in front of their parishes after Sunday mass. However, only three of the 18 priests (or 17 per cent) were assigned penances.[39] Charges against laymen were often resolved immediately: David Millewarde was accused of fornication with Isabel Mathowe on 30 May

1468, appeared in court the same day, confessed to the charge and was assigned a penance of two floggings. Charges against priests, however, were often simply never resolved. Although a charge of incontinence was also brought against chaplain Thomas of Bockleton and his concubine, Agnes, on 30 May 1468, he was neither summoned to nor appeared in court, and it seems as though the charge against him was never pursued.[40]

Not only were priests punished less frequently than their lay neighbors, but they were also more often allowed to fulfill their penances in private, thereby avoiding the public humiliation of a flogging. In 1468–69, two out of the 80 laymen who received punishment (3 per cent) were allowed to perform their penances privately by making an offering or a pilgrimage to Hereford Cathedral. In contrast, all three priests who received punishments in 1468–69 were assigned a private penance. Lewis ap Hopkyn, the curate of Mainstone, was warned to stay away from his concubine and assigned a penance of fasting on bread and water for a week, making a pilgrimage to the shrine of St. Thomas in Hereford Cathedral and reciting the nocturnal office every Sunday between matins and mass for seven weeks while standing bare-headed and barefoot in front of his church font.[41] Although elaborate, Lewis' punishment would have spared him having his sin announced in front of his parishioners at Sunday mass and walking in the penitential procession. Church court officers, it seems, tried to avoid damaging the authority of parish clergy with the public spectacle of a flogging.

Even as they abandoned celibacy and adopted some norms of secular masculinity, incontinent priests and lenient ecclesiastical officers still labored to maintain the authority and community standing of the clergy. Priests' *violation* of their vow of celibacy was public knowledge, but the *remedy* for their sin was only rarely a public spectacle. Whether court officers were treating their clerical colleagues more leniently than laymen, or were more concerned with reforming – rather than punishing – errant priests, or were simply trying to maintain the spiritual authority of the clergy is difficult to determine.

It is not difficult, however, to determine that these courts treated priests and their concubines in strikingly different ways that worked to reinforce the conventional masculinity of priests. Ruth Karras has pointed out that although there were many definitions of secular masculinity during the later Middle Ages, laymen – whether knights, university students or craftsmen – were made masculine by their heterosexual desire, their status as heads of household or their dominance over women.[42] So, too, were some of the priests of Hereford. Priests acted like laymen sexually and socially, and they also resembled laymen in another way: in their concubinary relationships, they held a privileged and dominant position over women. Not only were priests treated more leniently than laymen, they were also punished less severely than their lovers.

Although clerical concubines might have gained stability in a long-term relationship or through financial support, they faced significant disadvantages in the church courts and were punished more severely and more often than their partners. Although Thomas, the chaplain from Bockleton, was not punished, his concubine was. Agnes appeared in court, confessed to the crime, and was given a penance of seven public floggings.[43] While priests were often assigned penance they could perform in private, women were always given standard, publicly humiliating penances. Some women received harsh penances even when their partners were never formally summoned to court, or had even purged themselves of the charge. And unlike lay concubines, who might be encouraged or coerced into marrying their partners, concubines were frequently evicted from their homes.

The church court's most common way of dealing with clerical concubinage was to separate a couple. If they lived together, the judge might force a woman to leave their shared residence; if not, he might order her to leave the parish or even the diocese. In October 1486, for example, Thomas Lynke was summoned to court on a charge of fornication with Agnes Willyams, who lived with Thomas' sister. Agnes confessed the sin and was assigned a particularly harsh penance of eight floggings, which she did not fulfill. When she appeared in court a few months later, the judge offered to postpone her penance if she promised to leave the diocese within a week. But when the court reconvened later that year, she had not left, was still in a relationship with Thomas Lynke and was now pregnant. After being suspended from church services and then excommunicated, she finally left the diocese by the following February.[44] While a priest's status as a householder was fairly stable, especially if he held a benefice, his concubine might lose her home, economic support and partner quickly and unexpectedly. Priests stayed in their residences; their concubines were forced to leave.

Priests had sex like laymen; they established long-term relationships like married men; they wielded power and privilege over their lovers, as did most men. And sometimes, heterosexual desire eclipsed celibacy as a defining element of clerical identity. Although there is scant evidence of how ordinary men – parish clergy included – viewed their own masculinity, two priests have left evidence of how they perceived their sexual behavior. Richard Hall, the vicar of Leominster, openly flouted canon law on celibacy. In 1527 and 1528, Hall was summoned to court twice. In the first charge, he was accused of impregnating Joan Merycke; in the second, that he had a child by Elizabeth Joyner. Hall denied both charges and promised to disprove the accusations through the process of canonical compurgation (during which laymen and other clergy would swear to his oath of innocence), but he never did. By October 1530, Hall had still not fulfilled his compurgation and was again summoned to court to respond to the charges. He seems to have evaded court officers for another six months, but in April 1531, Hall was summoned and disciplined for preaching that a breach of clerical celibacy was a trivial offense. The court clerk considered Hall's actual words so crucial that he recorded them in English: 'Carnal pleasure', Hall proclaimed, 'is natural and it is not that thing that God takes vengeance for. For a little confession will easily remove it.'[45] Hall's lack of willingness or ability to find compurgators to swear to his innocence implies his guilt; his characterization of clerical fornication as a peccadillo makes clear that he did not take canon law on celibacy – or his own priestly vow – seriously. Some priests may have been torn between different models of masculinity, but Hall was a cleric whose sexuality was firmly a component of his clerical (and masculine) identity.[46]

Another Hereford cleric also took pride in his sexual misdeeds. Edmund Powell, a priest from the parish of Dixton, was charged in 1529 with impregnating a woman named Margaret. He appeared in court, denied the charge and was assigned a day to purge himself, but then declared that he had carnally known '100 women'. In reply, the judge immediately assigned him a penance of one flogging around the Hereford Cathedral. Powell's claim might well have been a deliberate exaggeration, if not an outright boast. Powell's choice of the number 100 was intended, I suspect, to convey an absurdly large or unimaginable number.[47] The number 100 may have functioned more widely as a rhetorical number, and it was sometimes used by court officers in just that way. In one case from 1509, Thomas Wilmottis, the vicar of Bridstow, confessed to having an ongoing relationship (and a child) with Maud Jeynkyns; the judge placed him under an obligation of 100s. if he committed the sin again.[48] Such a large fine would have seemed outrageous – at the time, most penances were commuted to 6s. 8d. – and it seems reasonable to assume that no court expected an individual to pay such a large fine. More likely, the fine was used for rhetorical effect, to emphasize the gravity of the sin committed by Thomas and Maud.[49] Powell's tally – whether it was meant literally, rhetorically or perhaps tauntingly – would have implied an exaggerated virility, a masculinity based on sexual bravado without regard for celibacy.

Taken together – and placed within the context of clerical fornication and concubinage – the claims of Hall and Powell suggest that, although celibacy was an essential marker of the priesthood in theory, it was not a practice of all priests, nor even a concern of some. Priests who flaunted their refusal to be celibate or boasted openly about their sexual virility were rare, to be sure. But they nonetheless show the possibility of a different conception of clerical masculinity, one that was far more conventionally masculine.

In recent years, the field of medieval masculinity has expanded greatly, particularly in terms of lay masculinity. No longer restricted to a simple tripartite model in which masculinity was equated with virility and manhood consisted of 'impregnating women, protecting dependents and serving as provider to one's family', historians now discuss diverse and nuanced models of lay masculinity that varied according to socio-economic class, life-cycle stage, marital status and occupation.[50]

Studies of the medieval clergy – perhaps too readily assuming that, in matters of clerical celibacy, practice followed prescription – have failed to take into account priests who did not live celibately. Yet some priests (and, perhaps, many priests) behaved both sexually and socially like laymen, participating in their communities as householders, fathers and quasi-husbands. Patricia Cullum has argued that, as clerical and lay life became increasingly similar in the later Middle Ages, distinctions between clerics and laymen were eroded due to the employment of clerics in lay occupations and the growth of the vernacular as an administrative language. Differences between priests and their parishioners must also have been elided when parish clergy, acting more like laymen than celibate clerics, took on aspects of secular masculinity.

Unchaste priests might have been a minority of the late medieval clergy, but they were not anomalies, and their presence invites a more flexible definition of clerical masculinity. Viewing clerics as a third gender is problematic because by removing priests from the dual-gender hierarchy, these theories obscure a basic element of both lay and clerical masculinity. It is especially in relation to women that we should see priests as men, not as an 'emasculine' gender, because – like laymen – they had power over women. As Ruth Karras reminds us, 'the subjection of women was always a part of masculinity', and studying unchaste priests might help elucidate what was quintessentially masculine in late medieval society.[51]

Acknowledgments

The research for this essay (and for the larger project of which it is a part) was made possible by two generous grants: a Schallek Fellowship from the Medieval Academy and the American Branch of the Richard III Society allowed me to spend 18 months in English archives, and a Mellon Foundation/ACLS Dissertation Completion Fellowship gave me the luxury of a year of writing without distractions.

Notes

1. Herefordshire Record Office (HRO), HD4/1/108, f. 182: *Dominus Rogerus Homme vicarius de Canon Frome incontinens est cum Isabella Herford uxore Jacobi Herford de Munsley quam tenet ex qua procreavit prolem ut ipsam duxit.* I have translated *duxit* as 'married' because in medieval Latin, the verb *duco* was specifically used in the phrase *duco in uxorem*, to take as a wife, with the sense of leading a woman to the doors of the church or to the altar.

2. For the Low Countries, see E.J.G. Lips, 'De Brabantse Geestelijkheid en de Andere Sekse', *Tijdschrift voor Geschiedenis* 102 (1989): 1–29; and Monique Vleeschouwers-Van Melkebeek, 'Mandatory Celibacy and Priestly Ministry in the Diocese of Tournai at the End of the Middle Ages', in *Peasants and Townsmen in Medieval Europe: Studia in Honorem Adriaan Verhulst*, ed. Jean-Marie Duvosquel and Erik Thoen (Ghent, 1995), 681–92. For France, see Raymond Eichman, 'The "Prêtres Concubinaires" of the Fabliaux', *Australian Journal of French Studies* 27 (1990): 207–13; and Jennifer D. Thibodeaux, 'Man of the Church or Man of the Village? Gender and the Parish Clergy in Medieval Normandy', *Gender & History* 18, no. 2 (2006): 380–99.. For central Europe, see Brigitte Rath, ' "De Sacramentis, Concubinatu et ludo taxillorum…": Über ein Böhmisches Visitationsprotokoll aus dem 14. Jahrhundert', in *Von Menschen und Ihren Zeichen*, ed. Ingrid Matschinegg, Rath and Barbara Schuh (Bielefeld,1990), 41–59; and Bernard Schimmelpfennig, '*Ex Fornicatione Nati*: Studies on the Position of Priests' Sons from the Twelfth to the Fourteenth Century', *Studies in Medieval and Renaissance History* 2 (1980): 3–50. For Spain and Iberia, see Pere Benito i Monclús, 'Le Clergé Paroissiale du Mareseme (Évêché de Barcelone) d'après les Visite Pastorales (1305–1447): Recherches sur le Thème du Concubinage', in *Le Clergé Rural dans l'Europe Médiévale et Moderne: Actes des XIIIe Journées Internationales d'Histoire de l'Abbaye de Flaran, 6–8 Septembre 1991*, ed. Pierre Bonnassie (Toulouse, 1995), 187–203; and M.A. Kelleher, ' "Like Man and Wife": Clerics' Concubines in the Diocese of Barcelona', *Journal of Medieval History* 28 (2002): 349–60. For the Italian peninsula, see Daniel E. Bornstein, 'Priests and Villagers in the Diocese of Cortona', *Ricerche Storiche* 27 (1997): 93–106.

3. Thomas More, 'A Dialogue Concerning Heresies', in *The Yale Edition of the Complete Works of St. Thomas More*, vol. 6, part 1, ed. Thomas M.C. Lawler, Germain Marc'Hadour and Richard C. Marius (New Haven, 1981), 295.

4. Ralph Houlbrooke, *Church Courts and the People During the English Reformation 1520–1570* (Oxford, 1979), 177–80; Tim Cooper, *The Last Generation of English Catholic Clergy: Parish Priests in the Diocese of Coventry and Lichfield in the Early Sixteenth Century* (Woodbridge, Suffolk, UK, 1999), 177; Margaret Bowker, *The Secular Clergy in the Diocese of Lincoln, 1495–1520* (Cambridge, 1968); Christopher Harper-Bill, *The Pre-Reformation Church in England, 1400–1530*, revised edition (New York, 1996); Peter Marshall, *The Catholic Priesthood and the English Reformation* (Oxford, 1994). For scholarship on clerical incontinence during the central middle ages, see C.N.L. Brooke, 'Gregorian Reform in Action: Clerical Marriage in England, 1050–1200', *Cambridge Historical Journal* 12 (1956): 1–21; Brooke, 'Married Men among the English Higher Clergy, 1066–1200', *Cambridge Historical Journal* 12, 2 (1956): 187–8; Julia Barrow, 'Clergy in the Diocese of Hereford in the Eleventh and Twelfth Centuries', *Anglo-Norman Studies* 26 (2004): 37–53; Barrow, 'Hereford Bishops and Married Clergy, c. 1130–1240', *Bulletin of the Institute of Historical Research* 60, no. 141 (1987): 1–8; and Brian Kemp, 'Hereditary Benefices in the Medieval English Church: A Herefordshire Example', *Bulletin of the Institute of Historical Research*, 43 (1970): 1–15.

5. Heath did not differentiate between fornication and concubinage (in fact, he did not mention concubinage at all in his analysis of the court book), and he assumed that 'only a small number of offenders eluded the vigilant detection processes of that age.' While Heath provided some examples of priests committing fornication or bequeathing property to their sons and daughters, he downplayed their significance, arguing that priests were simply 'easy' targets for accusations of sexual misconduct. Peter Heath, *English Parish Clergy on the Eve of the Reformation* (Toronto, 1969), 115–19.

6. It is telling that the one early (1976) study of clerical incontinence which showed a high rate of clerical incontinence has rarely been cited in more recent scholarship. Using records from the diocese of Chichester in 1506 and 1507, Stephen Lander found that 15 per cent of parishes in the archdeaconry of Chichester had an incontinent cleric. These findings from Chichester fall within the range of figures from continental Europe, but historians have tended to ignore them, perhaps because Lander's numbers do not conform to the standard narrative of England's chaste clergy. Stephen Lander, 'Church Courts and the Reformation in the Diocese of Chichester', in *Continuity and Change: Personnel and Administration of the Church of England 1500–1642*, ed. Rosemary O'Day and Felicity Heal (Leicester, 1976), 215–38.

7. Kathryn Kerby-Fulton, *Books Under Suspicion: Censorship and Tolerance of Revelatory Writing in Late Medieval England* (Notre Dame, IN, 2006). See also Dyan Elliott's comments on *Books Under Suspicion*: 'Comment: English Exceptionalism Reconsidered', *Journal of British Studies* 46 (2007): 753–7.

8. Robert Mills, 'The Signification of the Tonsure', in *Holiness and Masculinity in the Middle Ages*, ed. P.H. Cullum and Katherine J. Lewis (Cardiff, 2004), 109–26; Maureen C. Miller, 'Masculinity, Reform, and Clerical Culture: Narratives of Episcopal Holiness in the Gregorian Era', *Church History* 72 (2003): 25–52; Jacqueline Murray, 'Masculinizing Religious Life: Sexual Prowess, and Battle for Chastity and Monastic Identity', in *Holiness and Masculinity*, 24–42.

9. Murray, 'Masculinizing Religious Life', 25. See also the essays in that volume by Katherine Smith and Scott Wells for more on the deployment of military models for monastic life. For other work on clerical masculinity in medieval Europe, see John H. Arnold, 'The Labour of Continence: Masculinity and Clerical Virginity', in *Medieval Virginities*, ed. Anke Bernau et al. (Toronto, 2003), 102–18; Thibodeaux, 'Man of the Church' See also selected essays in the following collections: *Conflicted Identities and Multiple Masculinities: Men in the Medieval West*, ed. Jacqueline Murray (New York, 1999); *Masculinity in Medieval Europe*, ed. D.M. Hadley (New York, 1999); and *Medieval Purity and Piety: Essays on Medieval Clerical Celibacy and Religious Reform*, ed. Michael Frassetto (New York, 1998).

10. R.N. Swanson, 'Angels Incarnate: Clergy and Masculinity from Gregorian Reform to Reformation', in *Masculinity in Medieval Europe*, 160–77, quotation at 167; P.H. Cullum, 'Clergy, Masculinity and Transgression in Late Medieval England', in *Masculinity in Medieval Europe*, 178–96, quotation at 183. In other work, Cullum further explores the different life experiences of clerics and lay people, arguing that the clergy had a different life-cycle, due, in part, to their celibacy. Cullum, 'Life-Cycle and Life-Course in a Clerical and Celibate Milieu: Northern England in the Later Middle Ages', in *Time and Eternity: The Medieval Discourse*, ed. Gerhard Jaritz and Gerson Moreno-Riaño (Turnhout, 2003), 271–81. To date, only one English historian has addressed clerical sexual misbehavior and taken into account the effects of sexual behavior on clerical identity. In a recent article, Shannon McSheffrey has compared the attitudes toward unchaste clerics in a fifteenth-century conduct book to their treatment in the city's secular courts. The prosecution of priests' sexual sins, she argues, showed clerics' awkward place in their secular communities – they were embedded in their parishes, but also set apart from their lay neighbors, particularly because of their chastity. See Shanon McSheffrey, 'Whoring Priests and Godly Citizens: Law, Morality, and Clerical Sexual Misconduct in Late Medieval London', in *Local Identities in Late Medieval and Early Modern England*, ed. Normal L. Jones and Daniel Woolf (New York, 2007), 50–70, quotation at 63. I am grateful to Dr. McSheffrey for providing me with a copy of this article.

11. Cooper, *The Last Generation of English Catholic Clergy*, 77–80; R.N. Swanson, *Church and Society in Late Medieval England* (New York, 1989), 1–26; Swanson has argued that even in terms of administrative hierarchy 'no single diocese could ever be considered "typical" or "average", ' *Church and Society*, 24.

12. The diocese included parishes in Herefordshire, Gloucestershire, Worcestershire and Shropshire in England; and Montgomeryshire, Radnorshire and Monmouthshire in Wales. Hereford is also the largest town in Herefordshire. Throughout this essay, however, I use the name Hereford to designate the diocese, not the city.

13. In 1291, there were 316 parish churches in the diocese of Hereford. By comparison, the comparably sized diocese of Worcester had around 360 churches in 1291, and Norwich, one of the largest dioceses in medieval England, had over 1300 churches. Barrow, 'Clergy in the Diocese of Hereford'; *Taxatio Ecclesiastica Angliae et Walliae Auctoritate P. Nicholai IV, c. 1291* (London,

1802), 157–77. By the fourteenth century, the number of parishes in Hereford had risen to just under 400 parishes and chapels. William J. Dohar, *The Black Death and Pastoral Leadership: The Diocese of Hereford in the Fourteenth Century* (Philadelphia, 1995), 15; J. Caley and J. Hunter, *Valor Ecclesiasticus Tempore Henrici VIII*, vol. III (London, 1817), 1–48, 277–80.

14. J.C. Russell, *British Medieval Population* (Albuquerque, 1948), 306–13; Christopher Dyer and T.R. Slater, 'The Midlands', in *The Cambridge Urban History of Britain: Volume 1, 600–1540* (Cambridge, 2000), 626.

15. *Valor Ecclesiasticus*, vol. III, 1–48, 277–80; Gianetta Marie Hayes, 'Reforming the Frontier: The Clergy in Wales and the Diocese of Hereford, c. 1540–1640' (Ph.D. diss., Vanderbilt University, 2004), 50; Christopher Hill, *Economic Problems of the Church, From Archbishop Whitgift to the Long Parliament* (Oxford, 1956), 111. For a discussion of the drawbacks of the *Valor Ecclesiasticus*, particularly its tendency to overvalue incomes, see Cooper, *The Last Generation of English Catholic Clergy*, 78. Based on medieval lay subsidies, the county of Herefordshire had a fairly average level of aggregate wealth; see R.S. Schofield, 'The Geographical Distribution of Wealth in England, 1334–1649', *The Economic History Review* 18 (1965): 483–510, especially 503–9; and Dyer and Slater, 'The Midlands', 622–3.

16. The diocese of Coventry and Lichfield also contained Welsh parishes. Swanson, *Church and Society*, 2.

17. Kristine Rabberman discusses some aspects of ethnicity and cultural exchange in 'Marriage on the Boundaries: Cultural Contact and Marriage Formation on the Welsh/English Border, 1442–1526' (Ph.D. diss., University of Pennsylvania, 1998).

18. R.N. Swanson, 'Chaucer's Parson and Other Priests', *Studies in the Age of Chaucer* 13 (1991): 41–80, quotation at 64.

19. For scholarship on church courts and marriage, see Andrew Finch, '*Repulsa Uxore sua*: Marital Difficulties and Separation in the Later Middle Ages', *Continuity and Change* 8 (1993): 11–38; P.J.P. Goldberg, 'Gender and Matrimonial Litigation in the Church Courts in the Later Middle Ages: The Evidence of the Court of York', *Gender & History* 19 (2007): 43–59; R.H. Helmholz, *Marriage Litigation in Medieval England* (New York, 1974); Martin Ingram, 'Spousals Litigation in the English Ecclesiastical Courts c. 1350–c. 1640', in *Marriage and Society: Studies in the Social History of Marriage*, ed. R.B. Outhwaite (New York, 1982), 35–57; Shannon McSheffrey, *Marriage, Sex, and Civic Culture in Late Medieval London* (Philadelphia, 2006); L.R. Poos, 'The Heavy-Handed Marriage Counsellor: Regulating Marriage in Some Later-Medieval English Local Ecclesiastical-Court Jurisdictions', *American Journal of Legal History* 39 (1995): 291–309; and Michael M. Sheehan, 'The Formation and Stability of Marriage in Fourteenth-Century England: Evidence of an Ely Register', *Medieval Studies* 33 (1971): 228–63.

20. Bishop Trefnant's visitation is one of only a few extant medieval visitations, and one of the more detailed and complete manuscripts. Although it is widely used anecdotally, it has rarely been the focus of an extended study. For a description of the content of the visitation, see P.E.H. Hair, 'Defaults and Offences of Clergy and Laity, 1397', *Transactions of the Woolhope Naturalists' Field Club (TWNFC)* 47, no. 3 (1993): 318–50. I am grateful to Christopher Whittick of the East Sussex Record Office for providing me with a copy of his

translation of Hereford's visitation. I also consulted the original manuscript, held by the Hereford Cathedral Archives and Library (HCA), and deposited in the HRO, A1779. Arthur T. Bannister published an edition of the visitation, but it is riddled with mistakes and omissions: 'Visitation Returns of the Diocese of Hereford in 1397, Parts I and II', *The English Historical Review* 44 (1929): 279–89 and 444–53; 45 (1930): 92–101 and 444–63.

21. P.E.H. Hair, 'Mobility of Parochial Clergy in Hereford Diocese c. 1400', *TWNFC*, 43, part I (1979): 164–80. The poll-tax returns of 1377 list 824 clerics (329 beneficed and 495 unbeneficed clerics). In 1381, the poll tax returns show a total of 820 clerics (815 who held benefices and 65 who were unbeneficed), but this return may under-represent the number of poor, unbeneficed clerics, who did not hold a permanent benefice and were not taxed. Hair, however, estimated that there were 1000 secular clerics in the entire diocese, and I use this larger figure as a baseline, in hopes that it more accurately represents unbeneficed priests. Russell, *British Medieval Population*, 134–35; Hair, 'Mobility of Parochial Clergy', 170.

22. In my analysis of Hereford's clergy, I have counted as regular clergy only those men who lived within a monastic house. I have counted canons (both secular and regular) as secular clergy, because they lived within the parishes they served.

23. In the visitation, 463 charges of sexual misbehavior were made against laymen; 316 indicated fornication and 147 indicated concubinage. These figures do not include clandestine marriage, bigamy or abandonment, since those crimes did not have clerical equivalents. Of the 80 charges against secular clerics, 21 indicated concubinage; of the 21 charges against regular clerics, 1 indicated concubinage. Many men – both clerics and laymen – were charged with multiple offenses, so the total number of individuals charged differs from the total number of charges.

24. Vleeschouwers-Van Melkebeek, 'Mandatory Celibacy and Priestly Ministry', 681–92; Thibodeaux, 'Man of the Church', 396, fn. 10.

25. For a more detailed discussion of how these statistics from Hereford compare to Continental numbers, see Janelle Werner, 'Just As the Priests Have Their Wives': Priests and Concubines in England, 1375–1549' (unpublished PhD thesis, University of North Carolina at Chapel Hill, 2009), 205–6.

26. For a good introduction to English church courts, see Houlbrooke, *Church Courts and the People*, 21–54. Hereford's earliest extant act book dates from 1407; beginning in 1442, a series of act books (with some gaps) runs through the sixteenth century.

27. Contemporary sources, such as bishops' registers and secular court records, clearly show that many erring priests were either undetected or unprosecuted by the church courts. For example, in 1448, Katherine Cheyne was summoned to court for fostering lewdness between Robert Andreux, a chaplain, and Margaret Pyrdewy, but Andreux was never charged with fornication; HRO, HD4/1/90, f. 7.

28. I have chosen these four sample years because the act books are complete, include records from all deaneries in the diocese and have no missing or illegible pages. These yearly figures include all charges prosecuted during that year, including multiple charges against individual priests. I have tallied charges of spiritual incest under 'fornication' or 'concubinage', depending

on the type of relationship. The table includes charges brought against all clerics, both secular and regular, because most of the regular clerics summoned for incontinence were either monks serving as parish priests or canons who held parish benefices. For full statistics from Hereford's fifteenth-century court books, see Werner, 'Just As the Priests Have Their Wives', chapter 5.

29. HRO, HD4/1/94, ff. 30, 41, 48 (John Dyer and Lusot); HD4/1/108, ff. 2, 267 (John Ball and Katherine, Joan, Alice and Maud); HD4/1/94, ff. 55, 58, 61, 121 (Maurice and Maud Burton); HD4/1/113, f. 170 (Anna Decons and Christopher Wake); HD4/1/91, ff. 14, 15 (John Baker and Isabel Gentill).

30. HRO, HD4/1/113, f. 221 (Roger Parler and Joan Kiste); HD4/1/92, f. 16 (Nicholas and Cecilia).

31. HRO, HD4/1/89, ff. 108, 109, 110, 112, 113 and HD4/1/91, f. 136 (Thomas Nasshe and Joan Mawndfeld); HD 4/1/100, f. 134 and HD4/1/101, f. 90 (Lewis Hoptkyn and Dyddge); HD4/1/105, ff. 91, 101 (Owen ap Griffith and Alice). Based on the standard demographic practice of using a birth interval of 2.5 years, it seems likely that Owen and Alice's relationship had lasted at least 20 years, thus pre-dating his appointment to Meole Brace.

32. HRO, HD4/1/96, f. 60 (Richard and Alice); HD4/1/110, f. 246 (Thomas Wilmotts and Maud Jeynkyns); HD4/1/116, f. 108 (curate of Shrawardine); HD4/1/101, ff. 118, 129, 138 (Thomas Gogh and Margaret Bygolt).

33. HRO, HD4/1/108, f. 197 (John Davies); HD4/1/103, ff. 101, 103 (William Church and Margaret); HD4/1/94, f. 24 (Henry Mitchel and Agnes Hopkyn); HD4/1/101, f. 62 (John Carpynter and Margaret); HRO, MX 113 (Ralph Patten).

34. HRO, HD4/1/96, f. 16 (John Tewe and Dylly Dege); HD4/1/94, f. 126 (John Verne and Katherine); HD4/1/111, f. 223 (John Hore and Margery Tirrolde).

35. See Hair, 'Defaults and Offences of Clergy and Laity', 329–30; Bowker, *The Secular Clergy in the Diocese of Lincoln*, 117–18. For one example of a synod that warned priests against living with women, see the 1213–14 Statutes of Canterbury, which succinctly put it: 'It is not safe to live with women.' F.M. Powicke and C.R. Cheney, eds., *Councils and Synods, With Other Documents Relating to the English Church, vol. II, A.D. 1205–1313* (Oxford, 1964), 23–36, quotation at 26.

36. HRO, AL19/11, f. 14r.

37. HRO, AL19/11, ff. 25v–26r. Michaelmas, the 29th of September, was considered the end of the agricultural year in England and was also a common – though not universal – day to hire servants for the upcoming year. See Christopher Dyer, *Standards of Living in the Later Middle Ages: Social Change in England c. 1200–1520* (New York, 1989), 28, and Larry Poos, *A Rural Society After the Black Death: Essex 1350–1525* (Cambridge, 1991), 201. For a discussion of clerical households, see Cullum, 'Life-Cycle and Life-Course'. Cullum notes that the households of parish clergy, who often had female housekeepers or servants, might have resembled lay households. Even all-male clerical households, based on friendship or patronage, could replicate typical patterns of household formation in England.

38. HRO, HD4/1/94, ff. 139, 141. Floggings were typically performed on Sunday after mass, during which the penitent's sin had been announced. The offender – barefoot, dressed in the white linens and carrying a candle – would

process around the parish church up to six times (and, perhaps, as many times around the market), all the while being flogged by the parish priest or rural dean. This punishment was meant to be salutary (bringing grace to the penitent), but also to be a humiliating display of sin and a deterrent to other parishioners. Rosalind Hill, 'Public Penance: Some Problems of a Thirteenth-Century Bishop', *History* 36 (1951): 213–26; Houlbrooke, *Church Courts and the People*, 47.

39. HRO, HD4/1/94 (Weobley, Leominster, Ludlow and Clun).
40. HRO, HD4/1/94, f. 31. Isabel appeared in court the same day and also received two floggings; HD4/1/94, ff. 30, 34, 36. Many priests were ordered to evict their concubines but were never punished.
41. HRO, HD4/1/94, f. 71.
42. Ruth Mazo Karras, *From Boys to Men: Formations of Masculinity in Late Medieval Europe* (Philadelphia, 2003), see particularly chapter 1.
43. HRO, HD4/1/94, ff. 30, 34, 36.
44. HRO, HD4/1/102, ff. 167, 185; HD4/1/103, ff. 139–40, 149.
45. HRO, HD4/1/119, ff. 23r, 26v; HD4/1/121, ff. 33, 82: *Dominus Ricardus Hall vicarius de Leominster notatur predicare ut sequitur in anglicis* how carnall pleasure was naturall and yt was not that thynge that god taketh vengeans for a lytle confession wolde put it lyghtely away.
46. Hall's assertion might have reflected contemporary (and increasingly popular) Lutheran views of clerical celibacy, but he was also participating in a long-standing literary tradition that made light of illicit sexual behavior. Robert Mannyng, in his fourteenth-century penitential treatise, laments that the sin of lechery is not taken seriously: 'As some of these unlearned men say, / God of Heaven is so courteous / That he shall on Doomsday certainly / Forgive the sin of lechery / Lechery is but light sin – / He will have mercy of all therein.' In Idelle Sullens, ed., *Robert Mannyng of Brunne, Handlyng Synne* (Binghamton, NY, 1983), ll. 589–94.
47. HRO, HD4/1/120, f. 320: *In articulo illo vir comparuit negavitque articulum unde habet se purgare in xxvii die mensis Maii sua trina manu sed dixit quod centum mulieres cognovit carnaliter.* ... It is not too farfetched, I think, to compare Powell's boast to a more modern one: in his 1991 autobiography, Wilt Chamberlain claimed, 'If I had to count my sexual encounters, I would be closing in on twenty thousand women.' Chamberlain, *A View From Above* (New York, 1991), 251.
48. HRO, HD4/1/111, ff. 239–40, 261, 262, 264.
49. For a discussion of rhetorical numbers in ancient Greece and Rome, see Walter Scheidel, 'Finances, Figures, and Fiction', *Classical Quarterly* 46 (1996): 222–38.
50. Vern L. Bullough, 'On Being a Male in the Middle Ages', in *Medieval Masculinities: Regarding Men in the Middle Ages*, ed. Clare A. Lees, Thelma Fenster and Jo Ann McNamara (Minneapolis, 1994), 31–45, quotation at 34.
51. Karras, *From Boys to Men*, 11.

Part III

Clerical Masculinity: Contested Identities

8
Between Warrior and Priest: The Creation of a New Masculine Identity during the Crusades

Andrew Holt

That multiple notions of masculinity existed in medieval Europe can be demonstrated from Bernard of Clairvaux's early twelfth-century treatise promoting the Knights Templar.[1] Indeed, Bernard highlighted competing constructs of masculinity even within Europe's warrior class, as seen in his curious distinctions between the 'new knighthood' and secular knights. Bernard found it especially troublesome that secular knights, once the backbone of Europe's warrior class, had by his time grown to love 'effeminate tresses' and silk clothing decorated with silver, gold and precious stones that were impractical for combat. This was in stark contrast, Bernard noted, to the unwashed Templars, who wore their hair dirty and short and always maintained what, for Bernard, was the proper rugged appearance of the true warrior. Indeed, Bernard even took a jab at the highly prized traditional masculine persona of secular knights when, regarding their flamboyant apparel, he asked, 'Are these the trappings of a warrior or are they not rather the trinkets of a woman?'[2]

Bernard's concerns over proper masculine apparel were reflective of a much broader divide between secular and clerical notions of masculinity. Clerical culture had long promoted a very different standard of masculinity for its members than the one found among medieval society's warrior class. Indeed, clerics preached the virtues of humility, chastity and non-violence, ideals that contrasted with the values of a violent, flamboyant and often arrogant knightly class. Yet with the coming of the crusading era at the end of the eleventh century, neither of their competing masculine ideals was wholly sufficient in addressing the unique status of the crusader. Indeed, clerics argued that crusading was a form of holy war and so crusaders had to be holy to assure God's good-will, but it was also understood that crusading demanded brutal and bloody warfare, which was the domain of the brash and boastful warrior class whose actions had traditionally been equated with wrongdoing by the leadership of the Church.[3]

Reconciling the violence of the crusaders with their holy mission became a problem the Church had to address nearly from the inception of the crusading movement. In response, clerical authors developed a framework for a new type of masculine identity, specifically for crusaders, that was a hybrid of both traditional warrior and clerical masculine ideals. This new masculine identity was built on the theological and strategic premise that a holy warrior had to be pleasing to God to have any hope of success on the battlefield. This was to be accomplished, clerics argued, by the crusader humbling himself before God in manner, behavior and appearance. Indeed, the Crusader was to become a chaste and humble warrior, who did not celebrate his manly deeds through boasting of his achievements, but instead gave the credit to God. The traditional arrogance of the aristocratic knight, with his ostentatious dress and display of bravado would never be acceptable in wars waged for the cross; and no distractions, even the concerns of his wife, should impede his manly pursuit of serving God in a crusade. In some cases, clerics retained and promoted some of the traditional masculine features of secular warriors, but applied theological reasoning to modify or enhance them. A crusader's courage, for example, which was among the most prized of manly virtues, was supposed to be even greater than other warriors because he fought on behalf of God.

While medieval clerics seem to have had a great interest in the subject, the issue of crusader masculinity has been little explored by modern historians. Although the broader issue of gender is currently one of the more popular areas of historical study, there is, to date, only one major work that focuses exclusively on the issue of gender and the crusades.[4] The essays in this collection examine almost exclusively the subject of women and femininity in the Crusades. Very little scholarship has focused on a new type of crusader masculine identity or explored the nature of masculine gender with regard to the Crusades.[5] There are, to be sure, other scholars who have pointed to the 'hybrid' nature of the crusader as a cross between warrior and monk,[6] but none have drawn a similar conclusion concerning their newly constructed gendered identity as I suggest below.

I Knights and clerics: Competing masculinities before the crusades

Medieval warriors before the crusading era often had a tense relationship with the Church. Prior to the later eleventh century, and in contrast to later images of knighthood as an institution governed by chivalric ideals,[7] the knight was often a crude, brutish and violent figure whose behavior was rarely in accord with Christian principles.[8] Secular notions of masculinity provided knights and other warriors a framework in which their 'misbehavior' might be viewed as understandable, if not acceptable. On the key issue of sexuality, for example, engaging in sex with women was, from a social perspective, important for young members of the nobility from which the knightly class was drawn. This was the case if only for the reason of boasting to other men in what has been described as a 'common celebration of the physical subordination of women'.[9] It was also generally understood that a man's sex drive could become uncontrollable if provoked by the sexual allure of women.[10] In this sense, men were considered to be less at fault for sexual impropriety than the seductive women with whom they sinned.[11] Consequently, medieval European society was often more likely to excuse the bawdy behavior of their warrior class based on the assumption that the male sex drive was inherently susceptible to feminine charms and that sexual exploits with women were important for establishing a proper masculine identity in knightly society.

In contrast, the clerical reforms of the eleventh and twelfth centuries inspired a profound challenge to traditional notions of masculinity, especially regarding acceptable sexual behavior. The issue of sex, even within marriage, had long been a concern for medieval clerical authors and preachers, but this concern reached a high point during the Gregorian reform of the eleventh century.[12] Clerical authors increasingly argued that sex was unclean and defiled both body and soul, making it difficult to reconcile with the highest ideals of the Christian life.[13] These concerns about the sinfulness associated with sex provided the theological justification for the segregation of women from the clergy and were foundational to the rise of priestly celibacy in the eleventh and early twelfth centuries. Until this point, although monks were celibate, relatively little had been done to impose celibacy on the secular clergy. Yet in the eleventh century the reforming Church began to insist on its primacy over the secular state. To argue its primacy, the Church asserted its clergy was spiritually purer than the laity. Clerical celibacy became central to this claim as clerics, drawing inspiration from monastic culture,

increasingly emphasized the potential pitfalls of marriage and importance of chastity for spiritual purity.[14] The reformers efforts culminated
in a declaration at the First Lateran Council in 1123 that celibacy was
mandatory for the clergy.[15]

The transition to mandatory clerical celibacy was not without its problems. As Jo Ann McNamara has pointed out, 'An important class of
men institutionally barred from marriage raised inherently frightening
questions about masculinity. Can one be a man without deploying the
most obvious biological attributes of manhood? If a person does not act
like a man, is he a man?'[16] This dilemma was partly resolved, in clerical minds at least, by reaffirming the threat of feminine sexuality as
an impediment to the spirituality of men in general and the clergy in
particular.[17] Additionally, the heroic chastity of the clergy was in itself
considered a masculine attribute. To have discipline over one's body,
especially one's sexuality, in a world full of temptations, was a masculine virtue.[18] Indeed, women who had renounced their sexuality and
lived chastely were often considered to be 'manly'.[19] Finally, by virtue of
their ordination priests were understood to acquire a spiritual authority
and status that was clearly gendered as male.[20] This was perhaps made
most clear in their title and role as 'father' to believers. Thus the rise
of the celibate cleric in the twelfth century resulted in a new type of
clerical masculinity that was held as superior to traditional notions of
secular masculinity.[21]

II The ideal crusader: Humility, courage and a manly appearance

The new assertiveness and confidence of the reformed clergy embold-
ened them to seek important societal changes outside their ranks.
Perhaps most significantly, eleventh- and twelfth-century ecclesiastical
leaders sought no less than the appropriation of warfare for purposes
they deemed worthy or even holy.[22] The new emphasis on wars fought
for holy purposes found its greatest expression in the birth of the crusading movement. Crusades were perhaps the ultimate manifestation
of holy war as the combatants were technically pilgrims taking part in
no less than an armed penitential pilgrimage. Indeed, 'pilgrim' was the
common term throughout the twelfth century to describe those who
took part in crusades while 'crusader' did not become the normative
term until much later.[23] The anticipated hardships of the First Crusade
were promoted by clerical leaders as a type of redemptive suffering for
those formerly sinful knights who might now employ their skills with

God's favor. The cleric Guibert of Nogent, for example, noted that Pope Urban II, during his preaching of the First Crusade at Clermont in 1095, warned his listeners that previously they had 'waged wrongful wars' for which they 'deserved only eternal death and damnation'.[24] Yet with the calling of the First Crusade, the knightly class now had the opportunity to fight in 'battles which offer the gift of glorious martyrdom' and would earn them 'present and future praise'.[25]

The crusade, as a new type of holy war, required a new type of warrior, and the reformed clergy eagerly provided the framework. Those sinful knights seeking redemption through the crusade would have to take and maintain vows like those required of the humble and penitential pilgrim.[26] From the time they had taken their vow until its redemption, a process that could take several years, crusaders were required to avoid sin and receive the sacraments regularly. At all costs they were to avoid offending God on whose goodwill they depended for the success of the crusade. With the clergy dictating the framework for the new masculine identity of the crusader, it is not surprising that they rejected many traits commonly associated with less pious secular knights.

Humility had never been associated with the persona of brash or arrogant knights, who affirmed their masculinity in part by reveling in their status as warriors and boasting of their manly achievements.[27] Yet humility was, nevertheless, long considered a virtue of those in the service of God.[28] Because the clerical and warrior classes had always theoretically maintained separate masculine identities, the imposition of humility posed a conceptual problem for their competing masculine identities. Indeed, crusaders were, because of their vows, as much in service to God as the clergy, yet they were expected to act as warriors, for whom traditionally acceptable behavior was far different than that normally associated with the clergy. To resolve this dilemma, clerical authors argued that crusaders must embrace a new type of warrior identity; that of the humble warrior, who through his fear of God had abandoned the unseemly arrogance and haughtiness of Europe's traditional warrior class. The crusade was, after all, a penitential act, effectively an armed pilgrimage, in which the crusaders made atonement to God for their sins.[29]

Crusaders, influenced by a competing ideal of masculinity, were also expected to 'manfully triumph' in battle, but they were to avoid the excessive boasts of their secular counterparts.[30] Such a view was propagated by the clergy from the beginning of the crusading movement for even at Clermont in 1095 Pope Urban II stressed modesty (*modestum*) as one of the supreme Christian virtues.[31] According to the anonymous author of the *Gesta Francorum* the Pope told his listeners that they would be able to save their souls through the crusade only by 'humbly' taking up 'the way of the Lord'.[32] Indeed, according to the Pope, western Christians had little to be haughty about anyway in light of the poor state of the Holy Land. Urban argued that his listeners ought to be ashamed even to speak of the defilement of Christian places at the hands of Muslim conquerors.[33] He then contrasted the crusader ideals of humility and modesty with the 'great pride' of those who reveled in their status as knights while only killing other Christians in their pursuit of secular goals.[34]

The clerical concern over the necessity of humility for the ideal crusader continued throughout the crusading era. Bernard of Clairvaux, perhaps the dominant clerical voice of the twelfth century, also lashed out against arrogant knights in his 1130 work, *In Praise of the New Knighthood*. Bernard contrasted the humility of the crusading order with the arrogance of traditional knights by noting how among the Templars 'no arrogant word, no idle deed, no unrestrained laugh … is left uncorrected once it has been detected.'[35] Indeed, in the same work, Bernard condemned the proud and stressed the need for his readers to humble themselves.[36] Likewise, Pope Eugenius III, in his calling of the Second Crusade in 1145, emphasized the necessity of a 'contrite and humble heart' for crusaders seeking absolution for their sins.[37] So while knights may have traditionally boasted of their achievements as a means of affirming their manliness, the crusaders were ideally supposed to avoid such boasts and instead establish their manliness through only their deeds rather than their words.[38]

In conjunction with the ideal of humility, crusaders were to maintain a modest, although decidedly masculine, appearance. This was especially the case with the clothing they wore. Specifically, they were to avoid flowing sleeves and other dress that the clergy deemed both feminine and impractical for combat. Modest clothing was also more appropriate because crusaders were bound by the traditional requirement of chastity for pilgrims and, as a result, had little reason to maintain an attractive appearance during a crusade. The crusader ideal of modest dress was in stark contrast to the prevailing norms of the knightly class, which was often excessively concerned with being on the cutting edge of twelfth-century aristocratic fashion.[39]

Bernard of Clairvaux, for example, railed against secular knights who showed greater concern for their appearance than for their effectiveness in combat. He rebuked those who draped their horses in silk, painted their shields and saddles, plumed their armor with 'rags' and adorned their bits and spurs with gold, silver and precious stones and then, 'in all this pomp', charged to their deaths. Bernard argued that such adornments feminized knights who instead should exude the masculinity he expected from warriors. Indeed, Bernard condemned those who blinded themselves with 'effeminate tresses' and tripped themselves up with 'long, voluminous tunics' designed to bury their 'tender, delicate hands in cumbersome flowing sleeves'.[40]

For Bernard, in contrast to secular knights, the best warriors avoided going into battle on 'dappled and well-plumed horses' preferring instead strong and swift horses more suitable for combat. They also armed themselves with steel rather than gold, 'thus armed and not embellished, they strike fear rather than incite greed in the enemy'. On the 'rare occasions when they are not on duty' they spent their time repairing their 'worn armor and torn clothing'. and 'cut their hair short, cognizant that, according to the Apostle, it is shameful for a man to cultivate flowing locks'. They rarely even washed their hair as they preferred to 'let it appear tousled and dusty, darkened by chain mail and heat'.[41] Indeed, the ideal image of the Christian warrior for Bernard, in contrast to the prevailing pomp of the secular knighthood, was one that expressed humility and sacrifice through their humble and unkempt appearance, battle-worn armor and their use of practical weapons and horses.

Bernard's concerns were not unique. Indeed, clerical authorities throughout the crusading era held similar views. Pope Eugenius III, for example, in his calling of the Second Crusade, noted that those who fought on behalf of the Lord should not wear 'precious garments' or other things which 'proclaim lasciviousness'.[42] Eugenius warned that such concerns were a distraction from matters of arms, horses and 'other things with which they may fight the infidels'. Similiar views continued throughout the crusading era. The thirteenth-century preacher Guibert of Tournai, in a way reminiscent of Bernard, admonished secular knights for their love of 'riches, luxuries and honors' and unfavorably compared their dress with the humble attire of Christ, noting, 'Christ [was dressed] in poor cloth, you [wear] silk and finery; Christ rode on a donkey, you [ride] on fine horses with rich trappings;...Christ [wore] a crown of thorns, you [wear] hairpins, hats, bands, ribbons and garlands;...'[43] In contrast to such worldly knights, Guibert argued, '...the true sign of righteousness appears in the crusaders, who practice the service of God with their heart, mouth, and works...'[44]

A crusaders' courage in battle was, in keeping with traditional views of warrior masculinity, a decidedly masculine trait. Indeed, it had long been held in medieval Europe that combat was a 'manly' pursuit and the crusaders maintained this view.[45] Pope Urban, for example, had cited the success of the Franks' heroic ancestors, including Charlemagne and Louis, to spur his listeners to 'manly' deeds or achievements.[46] Pope Eugenius III also reminded crusaders of the necessity to 'gird themselves manfully' when battling infidels.[47] Such views among clerics continued throughout the crusading era. Bertrand de la Tour, for example, the thirteenth-century cardinal bishop of Tusculum, argued in his crusading sermons that warfare was 'manly combat' and dangerous 'without a strong arm'.[48]

One of the better examples to illustrate the association of masculinity with courage comes from an odd account of a clash between Christian and Muslim forces prior to the launching of the Third Crusade. In May 1187 a few hundred Templar and Hospitaller knights boldly attacked a Muslim force of seven thousand. Although the Christian forces were expectedly defeated, one Templar knight, Jakelin de Mailly, won tremendous respect from his Muslim opponents for his courage on the battlefield. Indeed, according to one account, after de Mailly had been slain, one of his Muslim opponents cut off his genitals and kept them in hopes of producing an heir 'with courage as great as his'.[49] Regardless of whether the story is true or not, the author's assumption that de Mailly's testicles would be recognizable to his readers as the source, or at least a symbol, of de Mailly's courage is reflective of the common association of extreme courage with hyper masculinity.

De Mailly's example was supportive of assumptions made by clerical authors that the crusader, in particular, had reason to be braver than any other warrior. After all, if they died while in service to God, they could be assured of a heavenly reward. Time and again clerical authors stressed how crusaders had nothing to fear from death as if they died in the service of a crusade they would die as martyrs. Secular knights and other warriors, on the other hand, did not have such assurances. If they died in combat there was no promise of 'martyrdom' or heavenly rewards as such benefits were reserved only for those who fought in the service of the faith. So while courage was expected of both crusaders and secular warriors, the crusader's courage was expected to be much greater than that of the secular warrior because he was assured of a divine reward for his sacrifice.

Such thinking was first made clear during Pope Urban's calling of the First Crusade in which he told his listeners that crusaders, in contrast to those who waged unjust wars, should have no fear of death as the crusade provided the opportunity for the 'glorious reward of martyrdom' and eternal praise for their sacrifice.[50] Bernard of Clairvaux also

emphasized how members of the crusading orders should be exceptionally brave in light of the spiritual protections accorded them. The ideal
Templar, for example, should be 'a fearless knight' who takes confidence
in knowing his 'soul is protected by the armor of faith just as his body is
protected by the armor of steel.'[51] For Bernard such a knight 'need fear
neither demons nor men' and nor should they fear death as when they
die they are finally able 'to be with Christ'.[52] In this sense, because of
the special nature of holy war, the crusader had good reason to be more
courageous than his secular counterpart and he was exhibiting a greater
masculinity when he did so.

III   The rejection of women and the maintenance of chastity

Crusaders were similar to clerics, as they too had made vows committing
themselves to the service of Christ. Crusades preachers eventually came
to understand crusading in terms of a *vocacio hominum ad crucem*.[53] In
this sense, crusading was a temporary vocation which began with a vow
that involved a spiritual reformation of the crusader's life and included
the imperative of chastity for both the married and unmarried.[54] It
was generally assumed that the married crusader, who effectively held
two vocations, would put his responsibilities to the crusade before his
responsibilities to his wife. In order to join a crusade, it was necessary
for a man to have the permission of his wife; yet this does not seem
to have been a major concern early on and many later preachers seem
not to have cared.[55] Consequently, there are prominent examples in crusade sermons of preachers praising men who disregarded the concerns
of wives and family members about going on a crusade. This rejection of
feminine authority came to be viewed by some clerics as a proper assertion of the crusader's masculinity. Indeed, in modern lingo, one might
say the ideal crusader was not 'henpecked'.

Beginning with the calling of the First Crusade, Pope Urban II urged
his listeners to avoid letting concerns about their families and possessions keep them from joining the crusade. He cited the Gospel of
Matthew in arguing that those who abandoned their wives, other family members and homes for the crusade would receive a hundred times
more in return along with everlasting life.[56] The Pope also made an effort
to alleviate more pragmatic concerns when he offered the family members of crusaders a degree of legal and financial protection through the
pronouncement of a 'horrible anathema' to all those who might 'dare
to harm the wives, sons, and possessions of those who took up God's

journey for all of the next three years'.[57] Consequently, the Pope's listeners had, in theory, no reason to resist the allure of the crusade as their families were protected at home. Such thinking was carried over into the later crusades, as clerics regularly warned crusaders against letting affection for their wives and families interfere with the business of crusading. Pope Eugenius III, for example, desiring to eliminate potential objections to taking the cross and citing Pope Urban's precedent, also decreed that the families of crusaders would be under the protection of the Church during the Second Crusade.[58]

Initially, the problem of men abandoning their wives and families for a crusade appears not to have been a great challenge, as generally there was considerable support among wives and families for male members joining a crusade. After all, women were also swept up in the widespread initial enthusiasm for the crusading movement and in some cases demonstrated even greater support for crusading than their husbands.[59] Yet following the failure of the Second Crusade, the support of women, as well as the general population, seems to have dropped off considerably. Consequently, clerics came to view women, particularly wives, as an impediment to their efforts to convince men to take crusading vows. In response, later crusades preachers increasingly argued that men who disregarded their wives' concerns about crusading were justified in doing so. The popular and influential crusades preacher James of Vitry, for example, maintained that it was good for a man to take crusading vows even if opposed by his wife. He illustrated this view through a colorful story contained in one of his surviving model sermons for the preaching of crusades. James lectured about a man who was forced to hide in his loft with his windows open so that he could listen to crusades preachers when they came to his town. According to James, the man could not attend these sermons in person because his wife objected to any notion of him taking the cross and locked him in the house when the crusade was preached in his town. Yet one day, while watching a preacher from his loft window, the man became 'inspired by God' and 'jumped out into the crowd and was the first to come to the cross'. James then celebrated the man's wisdom in defying his wife and listed the numerous spiritual benefits he would receive for doing so. Concerning the now defeated and abandoned wife, James disparagingly noted, '...the devil often extinguishes a good proposal through the spouse.'[60] Indeed, for James, and many other crusades preachers, this man's rejection of his wife's wishes and authority was a proper assertion of his own masculine prerogative.

Finally, crusaders, like all pilgrims, were to be chaste. As referenced earlier in this essay, this was a remarkable difference from traditional standards of acceptable behavior for the warrior class. Indeed, chastity was viewed as an essential part of the crusaders' identity and any question of sexual immorality during a crusade was likened to treason and resulted in swift and severe punishments.[61] Although chaste crusaders could not boast of their sexual exploits with women, as did their secular counterparts, their masculinity was never in question. The celibate clergy, after all, had already acted as trailblazers in this regard; they had successfully established that a man could be both celibate and masculine if done for spiritual reasons. It was also the case that events during the First Crusade seemed to confirm the necessity of sexual purity during a holy war for military victory, the greatest of manly achievements. One of the earliest and most significant associations of military defeat with sexual immorality, for example, took place during the lengthy and bloody siege of Antioch in 1098. At one point discouraged crusaders and clerics gathered together to discuss the reason it was taking them so long to conquer the city. They agreed that God must have been punishing them for their sins by prolonging their final victory. They came to the conclusion that to purify their forces, they needed to expel the women, both married and unmarried, from their camps to avoid 'the sordidness of riotous living'.[62]

After sending away the women, the fortunes of the crusaders did indeed seem to improve as a short time later they captured nearly all of the city of Antioch with only the exception of a well-defended citadel. Yet with the late arrival of a relief force of Turks the besiegers found themselves under siege as they struggled to defend their gains from a fresh and powerful Muslim army. In this predicament, the crusaders again searched for the source of their misfortune. After all, they had dismissed the 'sinful' women from their camps and were still having problems. Clerics again determined that immoral acts with sinful women were the cause of their misfortune. This time, they believed it was because crusaders had consorted with 'unlawful' local women. As a result of these sins, they believed God had 'doubled' their punishment.[63] Similarly, the anonymous author of the *Gesta Francorum* describes an incident at Antioch in which a priest had a vision of Jesus. Jesus complained to the priest that the crusaders' evil pleasures with Christian and pagan women were indeed the cause of their misfortunes.[64] Jesus imposed 5 days of prayer on the crusaders as a penance for their sins and, if this was done, he promised divine aid. Sometime later, in what must have seemed a confirmation of that promise, the repentant and obedient crusaders were victorious.

The surprising victory at Antioch seemed to vindicate those clerics who had argued that sexual immorality was indeed the cause of the crusaders' problems. After the battle for Antioch, instructions were sent to the West that all non-combatants, especially women, should remain at home.[65] A large number of clerical accounts appeared in the early twelfth century that emphasized that sins involving sexuality were the cause of any setbacks in the otherwise successful First Crusade. The cleric Raymond of Aguilers, for example, told the story of how the apostles Andrew and Peter appeared to Peter Bartholomew at Antioch to warn him that the Crusaders were having problems because of adultery.[66] Bartolf of Nangis cited an instance in which a man and woman were caught in the act of adultery and were publicly whipped to make atonement to God.[67] The priest Albert of Aix noted, '...the Lord is believed to have been against the pilgrim [crusader] who had sinned by excessive impurity and fornication.'[68]

Such thinking was firmly embraced by later clerics who wrote about the importance of chastity for success on a crusade. In the wake of the Second Crusade, for example, disillusioned supporters of the crusade sought accountability and answers for its failure.[69] Clerical responses ranged from those who claimed it had been indiscreet to allow women to join the crusaders to those who explicitly blamed the failure on the presence of women. The cleric Vincent of Prague, for example, claimed women were the source of sin and immorality in the crusader camps and cited them as the cause of defeat.[70] Other chroniclers, including Giselbert of Mons and William of Newburgh, echoed Vincent's sentiments.[71] Clerics were particularly concerned with the misconduct of Queen Eleanor of Aquitaine, who accompanied her husband Louis VII during the crusade, as they suspected her of adultery. Moreover, William of Newburgh was especially distraught that Eleanor seems to have inspired many other women to accompany their husbands at this development and later complained that the situation made the army unchaste and undisciplined.[72]

Clerical concerns about chastity and the presence of women that developed during the First and Second Crusades became the guiding principle of how clerics viewed women in subsequent crusades.[73] As a result of such concerns, several measures were taken in 1188 to restrict the number of female camp followers. The Councils of Geddington and Le Mans expressly forbade any woman except laundresses, who were deemed above suspicion, to go on the Third Crusade.[74] As a result, only elderly or unattractive women were permitted to participate.[75] Similar attempts to limit the presence of women during the Fourth Crusade included an effort by a papal legate to dismiss all women from the expedition while still in Venice.[76] Consequently, in the era of the Fourth Crusade, crusading was beginning to evolve from its pilgrimage roots into something more akin to typical military expeditions. As a result, fewer and fewer women participated. Only when the presence of women on a crusade was nearly eliminated did concerns over chastity become less of an issue for clerical authors of the crusades.[77]

IV Conclusion

The crusader was like no warrior Christendom had yet seen. He combined the most important virtues of the cleric and the knight in service to God and, in doing so, became a force that was respected by friend and foe alike. A new type of hybrid masculine identity emerged that embraced the traditional notions of warrior masculinity, such as extreme bravery and prowess on the battlefield, with a competing notion of clerical masculinity, in which humility, devotion and even chastity were upheld as the highest ideals of the holy warrior. This is not to imply that aristocratic notions of knighthood at odds with the Church did not continue to exist. Indeed, secular knights existed and maintained an ever-evolving view of masculinity that differed with the idealized notions of crusaders and the crusading orders well into the fourteenth and fifteenth centuries. Yet the crusaders and their offspring, the knights of the crusading orders, through strict clerical oversight, succeeded in creating a new masculine identity and established a new basis for judging the manliness of Christian warriors that challenged long-held secular notions of warrior masculinity.

Notes

1. In English translation, see Bernard of Clairvaux, *In Praise of the New Knighthood*, trans. M. Conrad Greenia (Kalamazoo, MI, 1977), 46. This translation is taken from the critical Latin edition prepared by Jean Leclerq and H.M. Rochais published in *Sancti Bernardi Opera*, vol. 3 (Rome, 1963). On the issue of multiple masculinities for groups as diverse as merchants, knights or priests, see *Medieval Masculinities: Regarding Men in the Middle Ages*, edited by Clare A. Lees (Minneapolis, 1994), *Conflicted Identities and Multiple Masculinities: Men in the Medieval West*, edited by Jacqueline Murray (New York, 1999), and *Masculinity in Medieval Europe*, edited by Dawn Hadley (New York, 1999).
2. Bernard of Clairvaux, 37; on the appearance of the Templars, see 46–7.

3. Ecclesiastical concerns about violence among the warrior classes were perhaps best expressed in the so-called Peace and Truce of God movements of the tenth and eleventh century. Through the Peace and Truce of God movements, the Church sought to limit violence within Latin Christendom by the imposition of various spiritual penalties on those who harmed non-combatants or fought on particular holy days. See H.E.J. Cowdrey, 'The Peace and the Truce of God in the Eleventh Century', *Past and Present* 46 (1970): 42–67, and *The Peace of God: Social Violence and Religious Response in France around the Year 1000*, edited by Thomas Head and Richard Landes (Ithaca, NY, 1992).

4. *Gendering the Crusades*, edited by Susan B. Edgington and Sarah Lambert (New York, 2002). There are several other works that examine the broader issue of women, and therefore issues of femininity, but there are no major works that focus on crusader masculinity. See, for example, Christoph T. Maier, 'The Roles of Women in the Crusade Movement: A Survey', *Journal of Medieval History* 30 (2004): 61–82, Rasa Mazeika, ' "Nowhere Was the Fragility of Their Sex Apparent": Women Warriors in the Baltic Crusade Chronicles', in *From Clermont to Jerusalem: The Crusades and Crusader Societies, 1095–1500*, edited by Alan V Murray (Turnhout, 1998), Helen Nicholson, 'Women on the Third Crusade', *Journal of Medieval History* 23:4 (1997): 335–49 and James M. Powell, 'The Role of Women in the Fifth Crusade', in *The Horns of Hattin: Procedings of the Second Conference of the Society for the Study of the Crusades and the Latin East*, edited by Benjamin Z. Kedar (London, 1992).

5. See Matthew Bennett's 'Virile Latins, Effeminate Greeks, and Strong Women: Gender Definitions of Crusade?' in *Gendering the Crusades*, edited by Edgington and Lambert (New York, 2002): 16–30.

6. See, for example, Malcolm Barber, 'Introduction', in *In Praise of the New Knighthood*, translated by Conrad Greenia (Kalamazoo, MI, 1977), 25. He notes, 'For the wider world the momentousness of the task which they had undertaken [the crusade] was surely justification for the creation of this new hybrid warrior-monk.' Yet the context of Barber's comment has nothing to do with masculinity.

7. Ruth Mazo Karras, *From Boys to Men: Formations of Masculinity in Late Medieval Europe* (Philadelphia, 2003), 22–3. Karras notes how until the twelfth century the knight (German- *ritter*, French- *chevalier* and Latin- *miles*) was essentially just a mounted warrior with little else to distinguish him in society. Not until the twelfth century and thirteenth centuries did the term 'knight' (or its equivalents) suggest somcone of means who lived according to a code of 'chivalry'.

8. Frances Gies, *The Knight in History* (New York, 1984), 4. See also, Peter R. Coss, *The Knight in Medieval England, 1000–1400* (Conshochen, PA, 1996), 46. Coss points out how medieval clerics viewed the knightly class as one driven by greed and violence.

9. Jo Ann McNamara, 'The *Herrenfrage*: The Restructuring of the Gender Systerm, 1050–1150', in *Medieval Masculinities* (Minneapolis, 1994), 8.

10. McNamara, 19.

11. Ruth Mazo Karras, *Sexuality in Medieval Europe* (New York, 2005), 121 and Jacques Rossiaud (cited also by Karras), *Medieval Prostitution*, translated by Lydia G. Cochrane (Oxford, 1988), 28. Women were often viewed as

temptresses that led men astray. Indeed, many medieval women internalized the blame for male desires as they felt it was their duty, not the man's, to insure they were not a cause of sin. See Karras, 39.

12. Jean Leclercq, *Monks on Marriage: A Twelfth Century View* (New York, 1982), 69 and P.J. Payer, 'Early Medieval Regulations Concerning Marital Sexual Regulations', *Journal of Medieval History*, 1 (1980): 370–1.

13. James Brundage, 'Prostitution, Miscegenation and Sexual Purity in the First Crusade', in *Crusade and Settlement*, edited by Peter W. Edbury (Cardiff, 1985), 57. These concerns were amplified by medieval notions of gender that held women had an insatiable sexual appetite while men had a nearly ungovernable sex drive. See McNamara, 19.

14. Karras, 43–4. This of course put enormous pressure on married priests, as well as those who held concubines, often *de facto* wives, to conform to the new and forceful call for a celibate clergy. Karras also notes 'Earlier, clerical marriage had been prohibited, but if priests married anyway the marriages were regarded as valid. From the twelfth century on, after clerical celibacy began to be enforced, many writers made a concerted effort to blame these women for polluting the Church and corrupting priests.' Karras also notes that priest's concubines were treated as the modern equivalent of 'gold diggers', out to despoil the priest of any possessions with their lusty sexuality. See Karras, 100–1.

15. Charles A. Frazee, 'The Origins of Clerical Celibacy in the Western Church', *Church History* 57 (1988): 126.

16. McNamara, 5.

17. McNamara, 8.

18. Karras, *Sexuality*, 37, 49.

19. McNamara, 6. See also the essay by Katherine Allen Smith, included in this volume, 'Spiritual Warriors in Citadels of Faith: Martial Rhetoric and Monastic Masculinity in the Long Twelfth Century.' Smith notes, '...the implication is that unlike their male counterparts, women who would be spiritual warriors must fight not only their evil impulses but their very nature. In the cloister, as in the secular sphere, a female warrior continued to be viewed as a powerful but unnatural figure. Many monks would likely have agreed with Gregory of Tours' assessment that, while they and their brethren fought 'as they should', women religious did so only by laying aside their feminine nature and, to paraphrase Paul, girding themselves with the armor of masculinity.'

20. Jennifer D. Thibodeaux, 'Man of the Church, or Man of the Village? Gender and the Parish Clergy in Medieval Normandy', *Gender and History* 18:2 (August, 2006): 383. Thibodeaux notes, '...reform era discourse...centered on the superiority of the 'godly virility' of the clergy over the 'lay virility' of secular men. To monasticise the secular clergy did not involve feminizing them; instead reformers adopted the rhetoric of clerical masculinity to persuade clerics to become a different sort of man.'

21. Thibodeaux, 381. Traditional prohibitions on bloodshed and the bearing of arms by clerics were also reaffirmed. Other behaviors traditionally associated with secular masculinity were rejected out of hand to include drunkenness, the frequenting of taverns, gambling and hunting. All of these

behaviors were formally prohibited in the Fourth Lateran Council of 1215. See Thibodeaux, 383.

22. The argument that the Church effectively sought to sanctify warfare in the eleventh and twelfth century was first made in the 1930s by Loren C. MacKinney, 'The People and Public Opinion in the Eleventh-Century Peace Movement', *Speculum* 3:2 (1930): 201, and Carl Erdmann, *The Origins of the Idea of the Crusade*, translated by M.W. Baldwin and Walter Goffart (Princeton, NJ, 1977). A more recent treatment of the issue is found in Tomaz Mastnak, *Crusading Peace: Christendom, the Muslim World, and Western Political Order* (Berkeley and Los Angeles, CA, 2002).

23. For an examination of the evolution of crusading and crusading vows as distinct from pilgrimage and pilgrimage vows, see James A. Brundage, *Medieval Canon Law and the Crusader* (Madison, 1969).

24. Guibert of Nogent, *The Deeds of God Through the Franks: Gesta Dei Per Francos*, translated by Robert Levine (Suffolk, 1997), 43.

25. Guibert of Nogent, 43. See also Fulcher of Chartres' account of Urban's speech.

26. Jonathan Riley-Smith, *The First Crusade and the Idea of Crusading* (Philadelphia, 1986), 23. Riley-Smith notes, '...in the twelfth century crusade and pilgrim vows were equated with one another and were, as far as we can tell, indistinguishable.'

27. Indeed, personal style and boasting of military achievements was an effective way for a knight to establish their manly reputation, and could sometimes build a knight's reputation just as well as actual campaigning. See Stephen Morillo, *Warfare Under the Anglo-Norman Kings, 1066–1135* (Woodbridge, Suffolk, UK, 1994), 45. See also pages 21 and 96 for additional commentary on the motivations behind knightly boasting.

28. For an overview tracing the development and necessity of humility for the clergy from St. Augustine to the High Middle Ages, see Julius Schwietering, 'The Origins of the Medieval Humility Formula', *PMLA* 69:5 (1954): 1279–1291.

29. According to Robert the Monk, Pope Urban II at Clermont (1095) referred to the crusade as a 'holy pilgrimage' (*sanctae peregrinationis*) and demanded the same vow as pilgrims. See Roberti Monachi, 'Historia Iherosolimitana', *Recueil des Historiens des Croisades, Occidentaux* 3 (Paris, 1866), 729, hereafter noted as RHC Occ.

30. Pope Eugenius III, 'Quantum Praedecessores', *Epistolae et privilegia* PL 180: 1065, 'viriliter triumphavit...'

31. Fulcherio Carnotensi, 'Historia Iherosolymitana', RHC Occ 3, 322.

32. *Gesta Francorum et Aliorum Hierosolimitanorum*, edited by Rosalind Hill (London, 1962), 1. 'ut is quis animam suam saluam facere uellet, non dubitaret humiliter uiam incipere Domini...'

33. Baldrici, *Episcopi Dolensis*, 'Historia Jerosolimitana', RHC Occ 4, 13.

34. Baldrici, *Episcopi Dolensis*, 14. 'Vos accincti cingulo militiae, magno superbitis supercilio; fratres vestros laniatis, atque inter vos dissecamini.'

35. Bernard of Clairvaux, 46.

36. Bernard of Clairvaux, 60.

37. Pope Eugenius III, PL 180: 1066. 'corde contrito et humiliato...'

38. If the sources are to be believed, Crusaders did exactly that, as one comes across numerous references to crusaders who give glory and often excessive thanks to God in the wake of their various successes. In the well known *Gesta Francorum et Aliorum Hierosolimita*, for example, which is one of the most important sources of the First Crusade, the anonymous knight who authored the text notes several such instances. In a typical example, he describes an incident in which he and other crusaders narrowly escaped from their Turkish attackers, noting 'Et nisi Dominus fuisset nobiscum in bello, et aliam cito nobis misisset aciem, nullus nostrorum euasisset, quia ab hora tertia usque in horam nonam perdurauit haec pugna. Sed omnipotens Deus pius et misericors qui non permisit suos milites perire, nec in minibus inimicorum incidere, festine nobis adiutorium misit.' See Rosalind M.T. Hill, *The Deeds of the Franks and the Other Pilgrims to Jerusalem* (London, 1962), 21.
39. Popular aristocratic literature of the era reflected popular fashions of the time. The anonymous author of the German epic *Nibelungenlied*, for example, described the hero Siegfried as a knight who 'dressed in eloquent clothes' which were made of silk and adorned with jewels and gold trimmings. See *The Nibelungenlied*, edited and translated by Arthur Thomas Hatto (Baltimore, 1965), 20–1.
40. Bernard of Clairvaux, 37.
41. Bernard of Clairvaux, 46–7.
42. Pope Eugenius III, PL 180: 1065. 'vestibus pretiosis…quae portendant lasciviam…'
43. Gilbert of Tournai, 'Sermo III', in Christoph T. Maier, *Crusade Propaganda and Ideology: Model Sermons for the Preaching of the Cross* (Cambridge, 2000), 205.
44. Maier, 205.
45. Megan McLaughlin, 'The Woman Warrior: Gender, Warfare, and Society in Medieval Europe', *Women's Studies* 17 (1990): 194.
46. Roberti Monachi, 728. 'Moveant vos et incitent animos vestros ad virilitatem gesta praedecessorum vestrorum.'
47. Pope Eugenius III, PL 180: 1065. 'viriliter accingantur…'
48. Bertrand de la Tour, 'Sermo II', in Christoph T. Maier, *Crusade Propaganda*, 238. 'Quia ergo patet campus ad congrediendum viriliter…periculosum sine forti brachio.'
49. This story is told in the *Itinerarium Peregrinorum et Gesta Regis Ricardi* and is translated by Helen J. Nicholson, *Chronicle of the Third Crusade* (Aldershot, 1997), 25–6. The story is also examined in Bennett, *Virile Latins, Effeminate Greeks, and Strong Women*, 16.
50. Guibert of Nogent, 43.
51. Bernard of Clairvaux, 34.
52. Bernard of Clairvaux, 34.
53. James M. Powell, *Anatomy of a Crusade: 1213–1221* (Philadelphia, 1986), 54. See chapter 3 ('The Vocation of the Cross') for an extended discussion of the vocation of crusading.
54. Complete sexual abstinence was expected of people doing penance, such as pilgrims. See Brundage, *Prostitution, Miscegenation, and Sexual Purity*, 57. The earliest crusaders understood themselves as pilgrims, albeit armed pilgrims, who through their suffering redeemed themselves from the effects of their sins. Because the crusades developed in the context of an armed pilgrimage,

the earliest crusading vows were essentially pilgrimage vows. It was not until around the year 1200 that church lawyers began to make clear distinctions between crusading vows and pilgrimage vows. For a further examination of the evolution of a crusade and crusading vows, as distinct from a pilgrimage and pilgrimage vows, see James A. Brundage, *Medieval Canon Law and the Crusader* (Madison, 1969).

55. One the formal requirement of the wife's permission to take crusading vows, see Riley-Smith, 22 or Constance M. Rousseau, 'Home Front and Battlefield: The Gendering of Papal Crusading Policy (1095–1221)', in *Gendering the Crusades*, edited by Susan B. Edgington and Sarah Lambert (New York, 2002), 33. For a primary source reference see Roberti Monachi, 729.

56. See Gospel of Matthew 19:29, 'And every one that hath forsaken houses, or brethren, or sisters, or father, or mother, or wife, or children, or lands, for my name's sake, shall receive an hundredfold, and shall inherit everlasting life.' Taken from the KJV. Regarding the pope's reference of this verse, see Roberti Monachi, 728.

57. Guibert of Nogent, 45.

58. Pope Eugenius III, PL 180: 1065.

59. Consider, for example, the case of Adela of Blois and her husband Stephen, who deserted his fellow crusaders at the siege of Antioch and returned home to much public scorn in Europe in 1101. Adela, concerned over the situation, persuaded her husband to return to the Holy Land to rejoin the crusade. See Orderic Vitalis, *The Ecclesiastical History of Orderic Vitalis*, volume 5, edited and translated by Marjorie Chibnall (New York, 1975), 324. This episode is also discussed in Rousseau, *Home Front and Battlefield*, 31.

60. See James of Vitry, 'Sermo II: Item sermo ad crucesignatos', in Maier, *Crusade Propaganda*, 120–3.

61. For the seminal work on this issue, see Brundage, *Prostitution, Miscegenation and Sexual Purity*.

62. Fulcherio Carnotensi, 340. 'Tunc facto deinde consilio, egecerunt feminas de exercita tam maritatas quam immaritatas, ne forte luxuriae sordibus inquinati Domino displicerent.'

63. Fulcherio Carnotensi, 345. 'Quibus visis, non minus solito iterum Franci sunt desolati, quia, propter peccata sua poena est eis duplicata.'

64. *Gesta Francorum*, 58.

65. Elizabeth Siberry, *Criticism of Crusading: 1095–1274* (Oxford, 1985), 44.

66. Raymond D' Aguilers, *Historia Francorum Qui Ceperunt Iherusalem*, translated by John Hugh Hill and Laurita L. Hill (Philadelphia, 1968), 76–7.

67. Siberry, 91. Siberry cites, Bartolf of Nangis, *Gesta Francorum Iherusalem expugnantium*, in RHC Occ 3, 498–9.

68. Alberti Aquensis, 'Historia Hierosolymitana' RHC Occ 4, 295. 'Hic manus Domini contra peregrinos esse creditur qui nimiis immunditiis et fornicario concubitu in conspectus ejus peccaverant …'

69. Numerous clerical writers and preachers who had advocated the crusade unexpectedly found themselves the target of sometimes vitriolic criticism. No less a figure than St. Bernard, perhaps the most popular and influential preacher of the twelfth-century, felt compelled to offer an apologia for his support of the crusade. See Bernard of Clairvaux, *De Consideratione Libri Quinque*, PL 182: 741–5.

70. See Vincent of Prague's *Annales*, MGH SS 27 and for Gislebert of Mons, see *Chronicon Hanoniense*, MGH SS 21.
71. The cleric William of Newburgh attributed the origins of the disastrous crusade, beginning with the fall of the crusader state of Edessa in 1144, to the lust of its naive governor, Jocelin II. According to William, the father of an Armenian girl who had been raped by Jocelin sought his revenge by betraying the city to the Turks and making possible Edessa's capture. See William of Newburgh, *The History of English Affairs: Book One*, edited and translated by P.G. Walsh and M.J. Kennedy (Wiltshire, England, 1988), 84–6.
72. Walsh and Kennedy, 128.
73. The clerical writer perhaps the most reflective of clerical views of women during the Third Crusade was Ralph Niger. He argued that the women who accompanied the crusaders served as the 'snare of the Devil' (Ecclesiastes 7:27) and worried that their presence might bring disaster to the Crusaders. To support his claims Niger cited the biblical example of the Midianite woman who was killed by the Israelites in order to prevent sin in their army and thus avoid defeat (Numbers 25:6–11). For such reasons, Niger argued that women should not be present on major military campaigns and that women should not participate in a crusade. See Ralph Niger, *De re militari et triplici peregrinationis Ierosolimitane*, edited by Ludwig Schmugge (Berlin, 1977), 227. See also 154 and 223.
74. Siberry, 45.
75. Keren Caspi-Reisfeld, 'Women Warriors during the Crusades, 1095–1254', in *Gendering the Crusades*, edited by Susan B. Edgington and Sarah Lambert (New York, 2002), 97. Caspi-Reisfeld argues that servant women who were old or unattractive were less of a sexual temptation to men. The twelfth-century poet/chronicler Ambroise describes this situation in his account of the Third Crusade. He notes, 'Because the women all refrained from going; in Acre they remained, save for the good old dames who toiled, and dames who washed the linen soiled and laved the heads of pilgrims – these were good as apes for picking flees.' See Ambroise, *The Crusade of Richard the Lionheart by Ambroise*, edited and translated by M.J. Hubert and J.L. La Monte (New York, 1941), 233.
76. Siberry, 46.
77. Although references to women, and by extension the threat of feminine sexuality to a crusaders' chastity, are considerably reduced in later sources, they do not disappear entirely. Indeed, during Louis IX's crusade to the East in 1248, Jean de Joinville recorded how the King dismissed a large number of crusaders believed to have interacted with prostitutes. When Joinville asked the King why he discharged so many crusaders, Louis responded by linking the 'debaucheries' of those crusaders with the 'suffering' and 'misery' of the army. See Jean de Joinville, 'The Life of St. Louis', in *Joinville and Villehardouin: Chronicles of the Crusades*, translated by Margaret Shaw (London, 1963), 207.

9
Knights, Bishops and Deer Parks: Episcopal Identity, Emasculation and Clerical Space in Medieval England

Andrew G. Miller

Just days before Thomas Becket, the archbishop of Canterbury, was murdered in Canterbury Cathedral by a band of King Henry II's knights (29 December 1170), he received word of outrages committed against him by members of the de Broc family, who were servants of the king and with whom the archbishop had been struggling over the control of the Canterbury estates since Becket's exile in 1167.[1] Among their offences, members of the de Broc family had intercepted one of the archbishop's ships carrying wine and killed or imprisoned the sailors; they broke into the archbishop's private park, slaughtered deer therein and seized his hunting dogs; they even cut off the tail of a horse in his service carrying household provisions – the mutilated beast was brought before Becket to see.[2] The archbishop's biographers report that such acts of violence steeled his resolve, preparing the way for manly resistance, holy defiance and blessed martyrdom.[3] William fitz Stephen, for example, attests that Becket 'was greatly strengthened in the Lord, behaving manfully and adorning himself with the armor of God, so that he might be able to stand valiantly in the day of the Lord; but he kept it secret, as much as he could, lest a disturbance arise in so festive [a season].'[4]

According to the chronicle accounts, the archbishop of Canterbury understood the message of violence, humiliation and emasculation that the knights communicated through invading his parks and killing and mutilating his animals. Certainly, Becket recognized that these acts were committed with the intention of defaming him publicly and in a disgraceful manner. Another chronicler reports that when news of the bloodshed reached Becket he exclaimed that 'a mare in my service has in contempt of my name had its tail cut off – as though I could be put to shame by the mutilation of a beast!'[5] Becket took the insult seriously enough to conclude Christmas Mass at Canterbury Cathedral by excommunicating those responsible for the violence; four days later the archbishop lay dead upon the floor of the cathedral, thanks in no small part to the de Brocs.[6]

Unsurprisingly, Becket's infamous execution by the knights of Henry II has overshadowed the significance of the insulting attacks perpetrated by the de Broc family, in particular the park invasion that preceded the archbishop's martyrdom. Still, both incidents reveal power struggles between secular and ecclesiastical authorities and, for our purposes, provide opportunities to explore the roles that private hunting grounds and prized animals played in conflicts between knights and bishops and what they reveal about both clerical masculinity and lay perceptions of it. To be sure, by entering Becket's parks with force, killing his deer, seizing his dogs and mutilating his horse, the de Brocs waged symbolic war against the archbishop within the confines of his private, masculine space, because his parks and his possessions represented secular elements of Becket's presence and influence as diocesan overlord. Private hunting grounds, then, served as an ideal venue for conflicts over competing definitions of masculinity and power in the Middle Ages; that is, the de Brocs invaded the archbishop's park – a well-understood symbol of a lord's masculine authority – in order to emasculate Becket publicly and expose his inability to protect his household and prized beasts.

This action is not the only example of powerful laymen shaming bishops through such forms of violent emasculation. A century later, in similar contests over power and authority, knights belonging to the household of Edmund, Earl of Cornwall (1272–1300) repeatedly broke into parks, slaughtered deer and mutilated animals belonging to the bishop of Exeter in an attempt to define local rights and liberties, in particular the creation and maintenance of private hunting grounds in the southwestern peninsula of England. Similarly, during struggles with the bishop of Lincoln over patronage rights in the diocese, household knights of King Edward I (1272–1307) butchered dozens of deer in a park belonging to the priory of Spalding and displayed the carcasses in a notorious and defaming manner. By overlaying royal records with the detailed episcopal *memoranda*[7] that survive from the registers of Bishops Walter Bronescombe of Exeter (1258–1280) and Oliver Sutton of Lincoln (1280–1299), it becomes clear that violence against an adversary's animals was comprehensible to both its lay perpetrators and clerical victims, and that the deer park was one of several venues in which gendered constructions of appropriate masculine behavior figured into contests over power and authority between knights, bishops and their respective households.

Recent scholarship has exposed the tensions and contradictions in medieval definitions of masculine behavior, demonstrating that there were many forms of medieval masculinities. Masculinity was directly related to one's profession and position in society. If secular masculinity in the Middle Ages can be defined as siring children, protecting dependents and providing for one's family,[8] then clerical, and in particular episcopal, masculinity might be defined similarly – with the obvious exclusion of sex – as marriage to the church, protecting dependents and providing for one's diocese and household. Therefore, when a knight flagrantly violated and desecrated a churchman's deer park and animals, he did so in order to scorn ecclesiastical power, rebuke and reject clerical expressions of masculinity, and draw attention to the clerics' ambiguous gender. The message of violence was all the more complicated in conflicts between bishops – who held both spiritual and *worldly* authority – and secular lords, since even knights had to reckon with an episcopal overlord's power and presence. The bishops' responses to the attacks against their parks and beasts accordingly reflected clerical (episcopal) understandings of masculinity and strength. While a knight might employ a sword and shield to assert his manliness, a bishop asserted his own masculinity by wielding the sword of anathema and shield of clerical immunity. In the same way, just as a knight might boast of his skills with women or weapons to emphasize his masculinity, a bishop might emphasize his own manliness by stressing his corporeal purity and the fortitude such sacrifice entailed. Indeed, it was precisely the spiritual and moral uprightness of the bishops that ultimately allowed these powerful churchmen to use their ecclesiastical powers to call the knights to task, fine them, and force them to seek penance for their invasion and slaughter, despite significant resistance and negotiation on the part of the knightly penitents.

Thomas Becket's biographers' portrayal of the archbishop's reaction to mortal peril is significant, revealing much about understandings of clerical masculinity in the High Middle Ages. For example, fitz Stephen emphasized the archbishop's calm, stoic reaction to the threats; Becket took the high road and played the greater man. The archbishop subverted the de Brocs' attempt to render him powerless and vulnerable to their violent offences. Instead of bowing in defeat or responding with physical retribution, Becket manfully accepted the threat to his own life, steeling himself in the face of looming martyrdom. Becket further demonstrated paternal concern for the public (his flock) by mastering his nerve and keeping the matter silent lest panic ensue from the impending violence. Clearly, the attacks against the archbishop's reputation, parks and property – and his reaction to them – offer insight into these competing notions of masculinity. The de Brocs' actions can be viewed as traditionally masculine (threats, use of force, violent humiliation and assertion of physical dominance) while Becket displayed another type of masculinity befitting his office as archbishop (episcopal masculinity), one that exemplifies the post-Gregorian Reform type of clerical hero cogently identified by Jo Ann McNamara and Jacqueline Murray.[9]

I Masculinity and hunting in medieval deer parks

While all levels of medieval society hunted or poached for sport and sustenance, scholars have long recognized that possessing conspicuous deer parks, and the prized venison they supplied for chase and banquet alike, served as prominent status symbols among the competitive English aristocracy.[10] Unlike the king's forest, chases or warrens, the deer park was a private, securely enclosed area;[11] it was small in comparison to a chase, yet was a prominent part of the lord's demesne lands. Wealthy landowners created such hunting grounds – which were incredibly costly – by imparking royal forest with the king's explicit permission. This privilege, called *rights of free-warren*, also allowed the recipients to hunt smaller game over their estates; the king frequently gave gifts of deer along with such licenses to help the beneficiary stock his park.[12] A luxury indicative of the new wealth of the High Middle Ages, the greatest expansion in the number of deer parks in medieval England was between 1200 and 1350. The greatest landholders – both lay and ecclesiastical – possessed numerous parks, ranging from a dozen to nearly 50. Because of the great expense and labor involved in constructing and maintaining them, deer parks were normally quite small in size, usually between 150 and 300 acres, though some were several miles in circuit; sometimes stretches of royal highway had to be entirely rebuilt, at the owner's expense, in order to circumnavigate the enclosures.[13]

Located near the manor house, castle or abbey and bound by massive earthen ramparts topped by a palisade of cleft wooden stakes for confining the deer and repelling trespassers, these private parks were physical and symbolic extensions of the lord's household and a reminder of his wealth and power.[14] The maintenance of the park and the welfare of the deer were a top priority since the deer provided the lord's household with a ready supply of fresh and salted venison and made possible the largesse – and accompanying cachet – of sharing the meat among friends and *familia*. As a result, deer maintenance was taken very seriously. For example, during the seasons of mating and fawning deer required greater privacy and protection, which often meant hiring extra people to keep the animals from being disturbed.[15]

While knights and their supporters broke into parks and slaughtered deer belonging to other powerful laymen – including the king – in medieval England, this type of violence was especially meaningful when directed against private hunting grounds belonging to bishops and abbots, for the right to hunt and impark forest – and especially the status indicative of this privilege – was a public expression of masculine power. In the Middle Ages, hunting was technically a pursuit unsuitable for churchmen. Like sex, hunting and hawking were forbidden to clerics by canon law because the use of weapons and mode of exercise were considered 'military'.[16] At issue between the secular and ecclesiastical authorities was control over deer parks as noble, masculine and secured space. Private hunting grounds were closely associated with domestic sanctuaries and were imagined as being spaces under similar means of control by virtue of one's own power or dominance; they were arenas highly charged with meaning, yet vulnerable to attack. Deer parks, then, were meaningful enclosures – semi-tamed corrals of wilderness – which, like fortified manors, served as aristocratic status symbols; their owners, lay and ecclesiastic alike, jealously safeguarded them. Indeed, because these spaces – symbols of a lord's masculine authority – were exposed, owners often found it necessary to employ armed parkers who, by the late thirteenth century, were licensed to kill poachers and intruders.[17] As a final point, it is important to consider that in England the fewest deer parks per square mile were located in the dioceses of Exeter and Lincoln.[18] Such a paucity of parks in these regions helped to make a man's possession of such conspicuous prizes all the more prestigious and, consequently, all the more contentious.

Invading parks and massacring or disfiguring deer was by no means uncommon in medieval England during masculine contests between powerful disputants; such violence proved provocative and meaningful for instigator, victim and audience alike.[19] For example, the de Brocs' violent incursion into Thomas Becket's private hunting space was part of a larger conflict between their household and the archbishop over controlling the Canterbury estates; subsequent archbishops continued to fall victim to knights invading their parks and massacring deer during other disputes.[20] Even medieval romances are devoted to the subject of defending one's parks, deer and reputation from violence and the resulting ignominy of a successful invasion. The English *Romance of Sir Degrevant*, written in the fifteenth century, depicts a manly contest between an earl and a knight who attempt to dominate and disparage one another by repeatedly invading each other's parks and slaughtering deer and swans.[21] To be sure, the invasion of an adversary's parks and the destruction of his animals – symbolic or physical – encoded well-understood conventions of violence, power and manliness; if they did not, it is improbable that the minstrel who composed *Sir Degrevant* would have spent so much of his romance graphically recounting park invasions and animal massacres to an audience he feared would be bored, disconcerted or perplexed by such tales.[22] In sum, to protect one's park and deer in the Middle Ages was to safeguard one's household, honor and manliness. To break into another man's park and massacre, his prized animals was consequently a highly provocative and menacing form of defamation and emasculation. Storming an enemy's private hunting grounds for sport and terror, although not as severe an act as invading his hall and slaying his retainers, exposed the owner's impotence as overlord, protector and provider – a grave insult which spoke to the essence of medieval manhood.

II Hunting and clerical manliness

Hunting was an act appropriate for worldly laymen; engaging in such sport clearly distinguished knights from monks and priests. Pursuing women and deer made knights more than male: it made them men. Especially among powerful aristocrats and their peers, the very act of hunting, as well as the number and quality of one's horses and dogs (let alone his parks), was a telling sign of his household's standing.[23] Hunting further served as preparation for and an avenue to worldly power. The chase conditioned knights for war, and hunting alongside one's king elevated a man's status and forged strong bonds between lord and vassal. Hunting deer in particular called to mind ancestral valor and noble privilege since the sport was, nominally at least, the preserve of the king and aristocracy.[24] A story from the late Carolingian period illustrates this association between hunting and lay masculinity. An uncle expressed his rage over his nephew's decision to become a monk instead of a secular cleric by demanding: 'How could you prefer the life of pigs in a vegetable-garden? What about the joys of hunting? What about the voluptuous touch of women?'[25]

Predictably, this prohibition against clerics engaging in the hunt was widely evaded by great churchmen in pre-Reformation England as clerics sought to engage in this manly pursuit.[26] Clerics were, on occasion, even guilty of poaching.[27] Moreover, medieval clerics petitioned for and claimed the right to hunt and impark forest – another potent symbol of masculine, worldly power.[28] Powerful English bishops and abbots possessed numerous parks, and royal law allowed them to take deer when passing through the king's forest.[29] Despite, or perhaps because of, this royal concession, deer parks belonging to prominent churchmen could serve as both focal points of contention with local noblemen and meaningful spaces for expressing grievances during disputes. For example, Matthew Paris describes a conflict between a band of knights and the abbey of St Albans during the 1240s and 1250s over local rights and liberties, including hunting. Paris vividly depicts Sir Geoffrey de Childwick astride his costly horse, surrounded by supporters, exhorting them to beat the abbot with staffs. Sir Geoffrey even let loose his pack of hunting dogs into the abbey's warren, followed by men who trapped deer with nets and shot them with bows and arrows. David Crook aptly observes that, for the monks of St Albans, a lord's privileges and hunting grounds were mutually exclusive.[30] Acquiring rights of free warren and especially practicing these rights had become an important symbol of the abbey's power and authority. Similarly, the integration of private hunting grounds into – and the power and authority they bestowed upon – ecclesiastical bodies is manifest in Jocelin of Brakelond's description of the installation of the abbot of Bury St Edmunds in the late twelfth century. The abbey's parks, which Brakelond prominently listed among its other properties, rights and liberties, clearly illustrate how these enclosures augmented the churchman's largesse and standing among his peers in the kingdom:

> Then the abbot had enquiries made in every manor that belonged to the abbacy about the annual rents paid by the free tenants, and the names of the unfree tenants, with their holdings and services. Everything was to be put down in writing. Furthermore, he repaired the manor-houses and domestic buildings that were so old and derelict that birds of prey and crows flew in and out of them. He built new chapels and added domestic apartments in many manors where previously there had been no buildings other than barns. He created

several parks which he stocked with game, and he retained a hunts-
man with hounds. If any important guest was being entertained, the
abbot would sit with his monks in a woodland clearing to watch the
hounds giving chase, but I never saw him eat the meat of hunted
animals. In his great concern to do everything to the advantage of
the abbacy, he had many pieces of land taken in from the wild and
ploughed up for cultivation, but I wish that he had taken a similar
care over the manors belonging to the convent.[31]

It seems clear that a park owner – even one who partook neither in the
chase nor in the fresh venison from his own deer – exhibited his power
and dominance by the very merits of possessing such a prized enclosure
and hosting great men within it. If the acquisition of a private, fortified
park in which to hunt deer at his leisure (or with his approval) was a
sign of the park owner's manliness and membership in, or aspiration to
join, the noble ranks, then a bishop's or abbot's elite hunting grounds
provided a type of arena – a masculine space under ecclesiastical dominion – in which laymen could communicate violently with clerics during
conflicts over clerical transgressions into traditional lay privileges and
pursuits.

III Episcopal masculinity in medieval Cornwall

The bishop of Exeter's creation of deer parks in Cornwall was central
to the violent disputes that erupted between the bishop and the young
new earl in the 1270s. Specifically, on 8 January 1259, when Edmund
was about 9 years old,[32] King Henry III granted Bishop Bronescombe
rights of free-warren throughout all his demesne lands in the bishoprics of Chichester, Exeter and Winchester.[33] The bishop subsequently
enclosed or took possession of parks at Cargoll, Paignton, Pawton, Penryn and Lanner, near St Allen. Significantly, Edmund's father, Richard,
Earl of Cornwall (1227–1272), had hitherto held a monopoly on hunting and imparking in the county.[34] Importantly, recent scholarship
argues that appearance was an important consideration for deer parks
possessed by ecclesiastical bodies and that sometimes the best land
available was imparked to achieve such aesthetics.[35] Indeed, medieval
patrons decidedly planned their deer parks and often intended to
emphasize particular features of the countryside in order to enhance
their status and fashion an eye-catching, elite setting.[36] It appears that
Bishop Bronescombe was a typical practitioner of this brand of park

aesthetic, since he used his new parks to publicize his power and authority throughout the region in precisely such a manner.[37] The bishop's
desire to exercise his new rights of free warren and create prominent
parks to augment his standing in the diocese/earldom helps to explain
why his secular male counterpart, the new earl, was so eager to attack
the bishop's parks – via prominent members of his household – upon
inheriting the earldom in 1272, at the age of 22.

Edmund was also eager to attack the bishop of Exeter's recently
constructed parks for another reason: the young earl was almost certainly a keen hunter, much like his first cousin, King Edward I.[38] There
were eight well-stocked deer parks in the county of Cornwall during
Edmund's reign. Restormel, which contained by far the largest park in
the southwest of England, was Edmund's favorite residence in Cornwall; he even made Restormel his home during his later years.[39] The
enjoyment and prestige that the earl gained from maintaining and
expanding his numerous parks outweighed the immense monetary
costs.[40] It is remarkable, considering Bishop Bronescombe's relatively
civil relations with Edmund's father, Richard,[41] that almost immediately
after Edmund became earl in April, 1272, members of his household
overtly violated the bishop's churches, parks, animals and clerics.[42]
The rights of free warren, imparking land to hunt, and the corresponding status these rights bestowed fomented strife and resistance
between the region's two great men and their households: one clerical,
one lay.

On 30 August 1272 – while both the bishop of Exeter and new Earl
of Cornwall were preparing (separately) to meet the king abroad[43] –
Sir John de Beaupré, the powerful sheriff of Cornwall who had been
hand-picked by the earl,[44] with a multitude of knightly and non-noble
co-conspirators alike, violently overran the bishop's parks at Cargoll
and Paignton, broke the enclosures, laid waste therein, killed deer and
attacked a number of household clerics who were present.[45] While the
king's records are vague about what transpired in the bishop's parks,
stating only that certain persons had violated Bronescombe and his
retainers by seizing their animals and other goods and breaking his parks
at Cargol and Paignton and taking away deer, the bishop's *memoranda*
paints a more graphic picture of the violent assault against his hunting
grounds – and especially the laymen's gruesome response to the bishop's
sentence of excommunication which he had subsequently levied against
the earl's men for their crimes. The bishop insists that the earl's men
also assaulted his clerics, laid waste to his parks and committed other
unspecified outrages.[46] One of the earl's knights later confessed in a

written statement that he had violated ecclesiastical liberty by climbing over the walls of the bishop's park at Paignton with his men and killed deer, while the earl's sheriff, Sir John de Beaupré, acknowledged that he had participated in the violence and shaming of the bishop's priests and clerks at Cargoll, and had ordered the enclosures of the park to be demolished.[47]

Clearly, these planned, large-scale and violent attacks against the bishop's private hunting grounds constituted something far more sinister than poaching deer. The earl's household – by breaking into episcopal parks, killing the bishop's deer and terrorizing his clerics – demonstrated to the people of Cornwall that Bronescombe was incapable of safeguarding his household and hunting grounds. Just as the de Broc family had shamed Thomas Becket by invading his park, slaying deer and seizing his hunting dogs, the earl's household waged symbolic war against the bishop of Exeter, emasculating and shaming him in a provocative, public manner.

The earl's men targeted Cargoll and Paignton because they were both prominent manors of the bishop. Bronescombe had recently acquired Cargoll from a prominent Cornish lord, and he often resided there.[48] The parks adjoining these episcopal manors, consequently, were meaningful targets for the earl's household to attack in order to convey the message of violence and mastery in the county, as well as to undermine the bishop's authority in the diocese. This brutal, publicized message of terror was not just limited to the bishop's hunting grounds. When the bishop ordered a number of clerics to warn the earl's household to desist from their armed conspiracy against the church of Exeter and its liberties, it provoked a decidedly macabre and defiant response from the parks' invaders. The episcopal register records how a band of the earl's men used swords and other weapons to beat the clerics, some of whom were cowering before the altars of the parish church. They then dragged the clerics around the churchyard with their horses. Furthermore, the earl's men defaced the clerics' clothing by cutting off the points of their cowls and hoods – apparently to mock the tonsure and ecclesiastical authority – and gruesomely mutilated their horses' tails, testicles, ears and upper lips, in order to humiliate and enfeeble the clergymen. This trenchant message of brutality was conveyed via the clerics' rough handling and the maiming of their helpless beasts to their episcopal overlord as well. Reminiscent of members of the de Broc family invading Thomas Becket's deer parks and mutilating a horse in the archbishop's service a century earlier in order to disgrace the archbishop publicly, the new Earl of Cornwall sent members of his household to terrorize parks belonging to Bishop Bronescombe and disfigure horses in his service in order to demonstrate in no uncertain terms that the earl's household was a formidable power in the region. The message of violence, mastery and emasculation was clear; the ruler of the *county* of Cornwall was superior to that of the *diocese* of Exeter. The earl and his household were – by force and by arms – the peninsula's greater men.

IV   Episcopal masculinity in medieval Lincoln

Although there is no evidence that the Earl of Cornwall's men either mutilated the bishop of Exeter's deer or displayed their carcasses in a provocative and shameful manner while invading the episcopal parks at Paignton and Cargoll, similar disputes in medieval England reveal that laymen also subjected deer to gruesome and consequential acts of violent disfigurement in order to embarrass and emasculate the diocesan overlord publicly and disgracefully. Twenty years later and over 200 miles to the northeast, Bishop Sutton of Lincoln was locked in a fierce dispute with a prominent household knight of King Edward I, Sir John St John,[49] over the right to present nominees of their choice to an ecclesiastical benefice: in this case the lucrative prebend of Thame, located just east of Oxford. (It is important to note that Sir John St John, as well as other knights in Sir John's household, personally knew Edmund, Earl of Cornwall).[50] At the height of the conflict with the bishop of Lincoln over the prebend, Sir John and his supporters, with a band of 200 men, seized the church of Thame on 8 August 1293.[51] One of the knights involved in storming the church was Sir Thomas Paynel, a household knight of Sir John and a companion of Sir Ranulph Rye, another household knight of the king.[52] Sir Ranulph Rye, like Sir John St John, was no friend of the bishop of Lincoln, for he was also fighting with the bishop over rights of advowson pertaining to the parish church of Gosberton.[53] The dispute was a personal one: The Rye family's patronage rights over Gosberton church extended at least back to the episcopate of Hugh of Wells (1209–1235).[54] When Thomas de Rye, the church's rector since 1272, died, Sir Ranulph presented the church to William Langtoft on 14 November 1295.[55] A couple of weeks later Bishop Sutton came to Gosberton to preach to the parishioners in the church. Sir Ranulph Rye, Sir Thomas Paynel and their supporters greeted the episcopal entourage by attacking the church with siege equipment and shooting arrows at the bishop's party; some of Sutton's men suffered wounds.[56]

Sir Ranulph Rye, Sir Thomas Paynel and Sir John St John, like Edmund, Earl of Cornwall, were apparently keen hunters and, by noble standards, better men for it. King Edward I granted all of them rights of free warren and, on occasion, deer from the royal forest. The king even allowed Sir John to create a 100-acre deer park on one of his manors in 1292.[57] When Sir Ranulph inherited his lands in the diocese of Lincoln in 1281, the king granted him and his heirs a weekly market and a yearly fair at his manor at Gosberton, as well as the coveted rights of free-warren in all his demesne lands.[58] Sir Ranulph liked to hunt so much that, just a couple of months after he and Sir Thomas Paynel had attacked the bishop of Lincoln at Gosberton, he and several other men were imprisoned at Nottingham for trespass in Sherwood Forest; however, Sir Ranulph and his cohorts were soon released on bail, despite their crimes, because the king needed them to fight in Scotland.[59]

Sir Ranulph Rye's church of Gosberton was literally next door to Pinchbeck, where a new deer park belonging to the prior of Spalding was situated.[60] This region of England, the Fens, was largely marshland; as a result many monastic houses sought isolation in the region. Draining and clearing marshland in order to make use of it was laborious and expensive; subsequently, land was at a premium.[61] Creating a deer park in the Fens was consequently a most conspicuous demonstration of wealth and power in a region with few great landlords and little available arable land. The priory of Spalding was a strong ecclesiastical presence in the region and, as the Benedictine priory became richer and more powerful in the course of the thirteenth century, the priors aggressively laid claim to estates in the vills of Weston, Spalding, Moulton and Pinchbeck. Importantly, the priory also sought rights of free warren on these lands, which King Henry III granted in 1236.[62] Tensions with the priory grew over course of the thirteenth century, stoked in part by the prior; he allegedly abused the limits of free warren on his manors from 1253–1274.[63]

On Thursday, 19 March 1293 – and for the second time during the course of the dispute – the bishop excommunicated the St Johns' supporters for savagely seizing the church of Thame.[64] Just as the Earl of Cornwall's men had violently interrupted the sentence of excommunication levied against them 20 years earlier by the bishop of Exeter, the household knights of the king and Sir John responded to the bishop of Lincoln's public defamation with a gruesome attack against the prior of Spalding's deer park in Pinchbeck. According to the bishop's register, on Easter Eve (28 March 1293), some unknown 'transgressors of divine and human law' (but undoubtedly the priory's troublesome neighbor,

Sir Ranulph Rye, and his knightly companions like Sir Thomas Paynel) broke into the prior's private hunting grounds and butchered nearly 60 deer 'by means of dogs and other cunning and not without malicious forethought'. Bishop Sutton was horrified and enraged by the brutality and extent of this humiliating assault against ecclesiastical authority in the region. Indeed, adding insult to injury, the invaders had displayed the beasts' carcasses in shameful or degrading positions. The bishop chose to depict the carnage in (anti)religious terms; he insisted that a demonic, sacrificial rite had been performed with the animals' blood: 'in insolence of the Host, and to the irreverence of that holy season, they immolated an abominable sacrifice to the devil in the blood of beasts, whereby a great scandal is said to have been stirred up in the aforesaid region.'[65]

Scholars have typically overlooked the significance of, and meaning behind, laymen invading deer parks belonging to prominent ecclesiastics in medieval England.[66] Instead of downplaying the knights' premeditated park invasions, it is more informative to begin by evaluating such publicized violence in light of scholarship which makes the connection between poaching and masculinity in medieval England and Normandy. First, poaching deer – let alone massacring dozens of them and displaying their carcasses in provocative positions (see below) – served to strengthen local and familial loyalties as well as to communicate with a rival by shaming or emasculating him. While hungry or opportunistic peasants poached deer in order to feed their families or to supply a lively (illegal) venison trade, noblemen most commonly poached in forests and parks for sport, male bonding and fresh meat (often for a feast). These knights preferred to pay steep fines – if they got caught – rather than be denied the thrill of the chase and the standing which resulted from a successful, heart-thumping foray. Expanding upon the work of Jean Birrell, Barbara Hanawalt has analyzed the role poaching played in exhibiting male identity and how besting the parkers guarding the royal forest was a form of masculine domination and defiance to the king's oppressive forest laws. Likewise, Jennifer Thibodeaux has explored how bands of poachers took deer and wood to assert masculine identity over the archbishop of Rouen during disputes over patronage and forest rights in thirteenth-century Normandy.[67] Birrell and others have also examined the ways in which poachers, armed and en masse, manhandled foresters during their raids in order to humiliate the guards as well as ridicule their overlord's authority and reputation, since it was the foresters' duty to defend their master's deer. Some gangs of poachers even seized a forester's hunting horn and weapons (symbols of his office) and left him in the wild to shame and emasculate him further and to help spread word around the locale of the poachers' manliness, defiance and camaraderie.[68]

What Sir Ranulph de Rye and his companions did at the prior of Spalding's new deer park in Pinchbeck was highly provocative and scandalous in the diocese of Lincoln for a couple of reasons. First, the attack against the priory's park in 1293 – like those attacks against the bishop of Exeter's parks at Cargoll and Paignton in the 1270s – was something far more sinister than poaching or manhandling the parkers; rather, the knights *havocked* the park at Pinchbeck in order to send a clear message of terror, ignominy and impotence to the prior of Spalding and his ecclesiastical overlord, the bishop of Lincoln. In Cornwall, the knights broke down the enclosures of the bishop of Exeter's parks, injured a number of clerics found at the scene and killed a number of deer.[69] Recall that invading an adversary's private hunting grounds was like storming his hall; such a public act of violence emasculated the park's owner as the household's overlord, protector and provider – wreaking such havoc attacked the heart of the victim's manhood. 'Havocking' is a military term denoting a command to despoil an enemy; it signifies far more than simply poaching or stealing deer. The word appears in Star Chamber grievances to depict poachers who killed more deer than they needed and left behind decaying carcasses. Breaking an adversary's parks and destroying his deer was a rebellious method that members of the gentry used in early modern England in order figuratively to 'slay' their opponents via massacring their prized beasts. Employing a hunting party as cover for a military raid or violent incursion against an opponent was a time-honored form of subterfuge, and one both practiced and recognized well into the sixteenth century.[70] Considering that medievalists have defined killing 16 deer at a time as an exceptional slaughter,[71] and slaying 42 in a day as astonishing,[72] and also bearing in mind that it took the queen of England and her sizeable entourage 2 days to kill a few dozen deer at Berkeley Castle in 1572, then slaying *60* deer within their guarded enclosures at Pinchbeck on Easter Eve 1293 constitutes a slaughter of extraordinary proportions.[73] Since camaraderie in the hunt and the taking of venison created strong bonds among the hunters, then ransacking an opponent's private hunting grounds and butchering dozens of deer in an act of violent retribution arguably forged even greater comradeship among the knightly participants; havocking deer in another man's park amounted to nothing less than a declaration of war. Finally, the intruders accomplished such a devilish feat with the help of hunting dogs and through 'other cunning', by which the bishop likely meant herding the deer into nets.[74] Using dogs to round up or attack deer was not just an effective means of slaughter, but the combined sound of horses and dogs also assured that the knights' assault constituted a grand performance and an eye-catching display of power.[75]

Second, it is significant that the knights chose to slaughter dozens of deer on Easter Eve, since Easter was the most important religious holiday in the Middle Ages. While it is true that the intruders did carry off a number of deer during their invasion at Pinchbeck[76] – and that noblemen preferred poaching excursions around Christmas, Easter and Whitsuntide – the massacre at Pinchbeck was particularly destructive and disgraceful because feudal lords were expected to offer munificence to their followers during this season with hunting and fresh game, for gifts of venison conferred masculine status in the medieval English household.[77] For example, during the fifteenth century, in order to make the most of his limited consumption of venison, the bishop of Salisbury served this delicacy only at major feasts.[78] By exacting their revenge against the bishop of Lincoln on Easter Eve, the knights slew two stags with one arrow, as it were: that is, the knights destroyed the priory's prized deer – which were living manifestations of ecclesiastical power, authority and largesse in the diocese – and, as a result, also violated the priory's sacred, masculine space 'against the wishes of the lords or guards'.[79] Sir Ranulph de Rye and his supporters struggled with the priory because it was a viable, meaningful target in a region in which their own families held property and influence. Through executing a 'diabolical' performance with the deer's carcasses, the knights upstaged the bishop's and priory's own performances of Easter Mass; the massacre at Pinchbeck had, by the bishop's own admission, 'stirred up a great scandal in the region'.[80] Reminiscent of the de Brocs invading the archbishop of Canterbury's parks and attacking his animals during the Christmas season in 1170, the knights of Edward I chose a prominent religious holiday in 1293 during which to shame their ecclesiastical opponents for maximum effect.

R. N. Swanson has noted that hostility against clerics in the Middle Ages was often depicted in terms analogous to those of medieval misogyny.[81] To be sure, during conflicts over church property and local privileges in late thirteenth-century England, knights and other laymen occasionally prevented clerics from performing a sentence of excommunication against them. Within the church itself, the disgruntled laymen disrupted the sacred performance, manhandled the churchmen before the parishioners and concluded their counter-performance by mutilating the ecclesiastical garments worn by the clerics (by cutting off their sleeves and the tips of their hoods and cowls) in order to empha-size the victims' religious identity while ridiculing episcopal power and influence. The humiliating assaults communicated – in a very public manner – the laymen's physical and symbolic mastery over the clerical 'body' and, perhaps more importantly, perverted the bishop's authority in the diocese. Similarly, laymen in medieval England occasionally cut off the tails of horses ridden by clerics acting on (arch)episcopal author-ity during local conflicts.[82] Resembling other types of anti-clericalism in the Middle Ages, then, the knights' violence against animals belonging to bishops and abbots in the dioceses of Exeter and Lincoln was a gen-dered response to perceived clerical or episcopal transgressions in their respective locales. Rather than slay the clerics, a crime that carried very serious lay and ecclesiastical penalties – not to mention public infamy – the knights instead invaded the churchmen's fortified hunting grounds, butchered their deer and in several cases placed the mutilated carcasses in provocative positions as an overt symbol of the bishop's inability to fulfill his masculine role as diocesan lord, protector and provider.

The knights' message of violence seems clear: the churchman (who should not be hunting in the first place) proved impotent even to defend animals within his personal, fortified, masculine hunting grounds. The knights' violent performance resulted in acute embarrassment, that is, the episcopal overlord was not nearly the paterfamilias he purported to be. By slaying the deer within the confines of their protected enclo-sures and by displaying the carcasses, the knights visibly exposed the churchman's inability to protect his animals which, when killed on his own terms, would have provided venison for his table or gifts for his friends and allies. The deer were especially meaningful victims for both the knights and their episcopal owners. First, the deer were wild, yet penned animals and, as such, were living symbols of masculinity and dominance. Like sex, hunting deer was a physical, violent under-taking not fit for those declaring religious purity and laying claim to the ecclesiastical privileges associated with a clerical lifestyle – even if all the churchmen did was profit from the largesse that the park and its perks (e.g. the abbot of Bury St Edmunds, above). Second, the mes-sage of violence transcended the animals themselves. The knights, by butchering the bishop's deer in such a grisly and provocative manner, communicated the terror of potential violence and mutilation against the churchmen and their episcopal overlord.

Although the bishop of Lincoln shrouded in religious terms what tran-spired during the outrage against the priory's deer at Pinchbeck, stating obscurely that the vandals had 'immolated an abominable sacrifice to the devil in the blood of beasts', the knights of King Edward I also used a number of the slain deer to shock the locale and generate lively discussion among the local inhabitants. What likely transpired was that the knights decapitated a number of deer and then stuck their heads either on poles inside the park itself or upon the spiked fence surrounding the hunting grounds. It was certainly not unheard of for a band of poachers to humiliate a park's owner, whether lay or ecclesiastical, in precisely such a manner. Without doubt, decapitating deer belonging to a powerful and authoritative man in medieval England was meaningful for a couple of reasons. First, decapitation was a well-understood convention in the Middle Ages, as was sticking a head – whether a human's or an animal's – upon a stick or pole to shame an adversary.[83] Second, performing this ritual mutilation with deer taken from a park or forest was distinctly shameful to the animal's owner because, according to the traditional rules of the hunt in the Middle Ages, decapitating the vanquished beast after an exhilarating chase marked the climax of a successful hunt. It was one thing, then, to poach another man's deer and take the venison either for sport or profit; it was quite another to cut off the head of the slain deer and display it in such a provocative and defaming manner, since the act deprived the lord of his symbolic prize and shamefully inverted the noble ritual of the hunt in which the victors would have paraded the deer's head before them upon their triumphant return.[84]

Prominent victims of such ignominious decapitation include the Earl of Lancaster, the father of Sir Thomas More, the king of England and the archbishop of Canterbury. This quartet of examples intriguingly suggests that when laymen invaded parks and forests belonging to another layman, they shamed him by cutting off the heads of male deer, while does were chosen when shaming the archbishop. To begin, on 23 March 1334, a band of 40 men, carrying bows and leading a pack of dogs, entered the Earl of Lancaster's Pickering Forest and killed 42 harts (male deer) and hinds (female deer); the poachers then impaled the heads of nine harts on stakes and left them as a gruesome symbol of contempt for the foresters.[85]

Similarly, when a gang of poachers killed deer belonging to Sir Thomas More's (1478–1535) father in response to his illegal imparking of land for a deer park, the hunters, as a ritual insult, impaled a buck's head upon a staff, placed a stick in its mouth and faced the grisly spectacle toward More's manor house.[86] A detailed account of this type of violent shaming survives from 1272, when a nobleman named Simon Tuluse, with a group of his household and neighbors, poached for several days in the king's forest and killed eight deer. On the last day of the hunt, after slaying three more deer, Tuluse decapitated a buck. He put the head 'on a stake in the middle of a certain clearing … placing in the mouth of the aforesaid head a certain spindle; and they made the mouth gape towards the sun, in a great contempt of the lord king and his foresters'. When the foresters saw this conspicuously insulting message and tried to attach Tuluse and his household, the latter attacked the foresters by shooting arrows at them and then carrying off the venison in a cart.[87] At the inquisition regarding Tuluse's crimes against the king, his foresters and deer, however, it was stated that the deer head had been a doe's, not a buck's, and that the object placed in its mouth was a billet (club or stick).

Barbara Hanawalt has rightly observed that, whether it was a spindle and buck or billet and doe, the insult was clearly one of sexual inversion; the buck had a spindle, or female symbol, placed in its throat while the doe had a club, or male (phallic) symbol.[88] Since the gendered meaning behind the insult seems clear and effective, regardless of which head and object combination was used, then why had the distinction become enough of an issue in the Forest Courts to merit not only clarification, but also valuable space in the records? Is it possible that Tuluse and his party were attempting to reduce or escape the charges by establishing in court that they had intended to shame someone other than the king – such as a churchman – by using a doe's head instead of a buck's? Indeed, in 1393 a certain Thomas Stodeman and his supporters came armed to the archbishop of Canterbury's parks at Ringmer and Mayfield, hunted without his license, killed deer and, rather than decapitating a buck or hart to defame this great lord, the laymen instead 'cut off the heads of *does* in time of venison and affixed them to the park palings, and assaulted his servant and parker John Henry'.[89] In 'time of venison' (*fermisona*) designated a closed season during which the deer were allowed to rest; hunting male deer took place between 24 June and 14 September while the season for female deer continued until 2 February.[90]

It appears that, because the act of chasing and killing deer forged bonds of masculinity between the participants – whether the men were hunting or poaching – knights and other laymen chose to subject lay owners of parks and forests to a more masculine form of ridicule, that is, by using the head of a male deer in the shaming ritual, while they subjected churchmen to a more emasculating form of ritual humiliation – desecrating the heads of does – in order to evoke the churchman's emasculation when it came to men's games like hunting deer. By killing does and placing their heads on the park's enclosure for all to see,

Stodeman's party wreaked even greater havoc by announcing to every-one in the region that the archbishop was incapable of protecting female deer even in their most vulnerable state. Finally, this insult might have proved particularly unsettling to this particular archbishop, William Courtenay (1381–1396), for his predecessor's own head had been cut off and stuck on London Bridge during the Peasants' Revolt of 1381.[91]

Historians who analyze the uses and meanings of violence in soci-ety – whether medieval or modern – agree upon one important point: destroying or mutilating an adversary's animal, whether a horse, deer, dog or swan, was not a crime against the beast but an expression of odium toward its owner; it was intimidation, terror and symbolic execution.[92] Contemporary audiences are not unacquainted with such provocative forms of brutality. For example, in Francis Ford Coppolla's *The Godfather* (1972), a producer is sent a powerful and grisly mes-sage after refusing to cast Johnny Fontaine in his movie; the gangsters placed the decapitated head of Johnny's favorite horse in his bed for him to discover. To be sure, violence against the animal is potential violence against the owner. In the Middle Ages, when powerful ele-ments of an owner's animal were removed – especially by means of an overt, humiliating ritual – the owner himself was publicly emasculated and symbolically rendered less powerful.[93] The violent slaughter and mutilation of the bishop's and prior's deer and horses were not crimes against the animals but against their ecclesiastical overlords; the ani-mals were property, and were havocked or disfigured in order to defame and ridicule their owners. After all, horses and deer were costly, cov-eted and dearly cared for in the Middle Ages; therefore, pitilessly killing or disfiguring valuable, status-evoking animals was not a random act of violence, but an action using encoded, well-understood conventions. Without these conventions, such acts would have seemed pointless to medieval audiences.[94]

V   Bishop takes knight: Excommunication, penance and masculine power

In order to use the ceremony of excommunication most effectively, the bishops had to turn the violent attacks against their parks, animals and reputation into decidedly religious issues – issues over which the churchmen had control of condemning and correcting as the regions' dominant spiritual authorities. Since these contests between knights and bishops over power and influence in their respective dioceses were played out in violent, gendered terms within the context of public performances (invasion of private deer parks) and counter-performances (rite of excommunication in a church), it is important to appreciate how legitimacy in medieval Europe was reliant on the communal sphere.

In medieval society, power was a reality that required continual reaffirmation by public performance and expression in order to direct the eyes and ears of the community to legitimize one's own authority or to undermine another's.[95] Carol Symes has examined a dispute from 1270 in Arras between laymen and clerics during which the count of Arras and his household struggled aggressively with the church over the city's public sphere. In the course of the conflict the laymen employed various tactics to publicize their supremacy in and around the city, such as violating sanctuary in the cathedral, spilling blood therein and choosing holy days to perform overt acts of violence and disruption against the interests and liberties of the church. On the Feast of the Assumption, bands of laymen broke into church buildings, seized animals and let a hunting falcon fly around during the Mass. In response, the laymen were excommunicated; some were ordered to make penitential pilgrimages while others were led in a procession to the cathedral on the Nativity of the Blessed Virgin, barefoot, dressed humbly and bearing rods in order to humiliate the penitents ritually and thereby achieve reconciliation in the community both materially and symbolically.[96] The clear pattern of competing lay and ecclesiastical performances, as well as the publicized contention over power, authority and legitimacy, helps to inform the present study in its examination of how and why knights and bishops asserted their respective masculinities during violent conflicts over imparking and patronage rights in medieval England.

Bishop Bronescombe of Exeter – through arbitration, arm-twisting and repeated condemnations and excommunication – eventually succeeded in asserting his power in the southwestern corner of England by acting as an episcopal overlord wielding the righteous sword of anathema against the earl's unrighteous invaders. Despite the grisly and public nature of the bold attacks, the bishop was acutely aware of his power's limitations in the region, for he made a point *not* to subject the earl himself to such anathema. As for the earl's knights, Bronescombe stuck to his mitre and waited them out; however, the bishop's sentence of greater excommunication continued to wear at the knights, so after scoffing irreverently at the penalty for years the knights came before the bishop and agreed to sign written statements ensuring their future good behavior.[97] Tellingly, in their written confessions to the bishop, the knights acquiesced that it was because they had feared for their souls that they had come to their senses under a renewal of divine providence and humbly and devoutly begged forgiveness from their episcopal 'father', 'lord bishop', and 'liege lord'.[98] The earl's household was then forced to seek absolution, repair the bishop's parks, restock them with deer and pay substantial fines. The bishop, by applying legal and religious pressure and by exerting his episcopal masculinity as the diocese's spiritual defender, emerged from the contest the greater man despite the tremendous cost in time and treasure.

Bishop Sutton of Lincoln – in his sentence of excommunication against the invaders who had laid waste to the prior of Spalding's park at Pinchbeck and massacred 60 of his deer on Easter Eve 1293 – depicted the slaughter as a diabolical attack against the Eucharist rather than a simple case of breaking the priory's park and slaughtering his living property in order to shame and emasculate the churchmen.[99] While it appears that the knights ultimately escaped punishment stemming from the invasion at Pinchbeck, the bishop was nevertheless successful not only in retaining the church of Thame but also in bringing Sir Ranulph de Rye to heel. Indeed, Bishop Sutton ordered Sir Ranulph de Rye and his supporters who had assaulted him at the church of Gosberton in 1295 to carry all the weapons and siege equipment which they had used to attack the church through the streets of Lincoln and into the cathedral in a penitential procession, and afterward to stand outside the church of Gosberton, displaying the armaments they had employed in the attack.[100] Although Sir Ranulph did not seem to care whether his followers were subjected to this sort of public humiliation, he himself refused to perform such unknightly penance.[101] The standoff dragged on for several months, until Sir Ranulph returned from fighting in Scotland on the king's behalf.[102] Bishop Sutton threatened to renew his sentence of excommunication if Sir Ranulph did not obey the bishop's commands. Eventually, Sir Ranulph appeared before Bishop Sutton in London, where they struck an agreement: the knight would not have to suffer the public humiliation of standing before the church doors, in plain view of the entire parish, but rather inside the building and with a squire carrying the weapons.[103] It was one thing for the bishop to excommunicate a knight, another to assign him penance and shame him and quite another to expect him to perform such chastisement publicly. Just as Bishop Sutton had made concessions with the St Johns' and king's household regarding punishments and reconciliations stemming from the attack at Thame, the bishop struck a similar face-saving deal with Sir Ranulph regarding the violence at Gosberton.

English bishops experienced limitations to their power and authority when it came to exercising their rights of free warren and excommunicating the knights who symbolically violated them by invading and wreaking havoc within their episcopal parks. Regardless of these limitations, what Bishops Bronescombe and Sutton lacked in the ability to punish prominent figures, man to man, they made up in the widespread and repeated denunciations of those whom they did excommunicate. Specifically, both bishops had to content themselves with humiliating and castigating lesser members of the aristocratic households for crimes organized by their betters. In turn, it was the very performance of such public punishments – by the bishops themselves or through their archdeacons and rural deans – that helped to confirm and reconfirm the bishop's masculine power and authority in the diocese.[104] A vivid example of how post-Conquest bishops used the command and coerciveness of excommunication to punish park invaders comes from a letter of Bishop Herbert Losinga of Norwich (1094–1119) written to local authorities and parishioners. While in Losinga's case poachers broke into the bishop's park at Homersfield at night and killed but a single deer – albeit the *only* deer in the park – and stole its carcass (leaving behind its head, feet and intestines), the bishop's incensed response is most informative. He begins by comparing a bishop's relationship with his flock to that of a bridegroom and bride and then sternly reminds the parishioners of their spiritual duty to divulge any information they may have about the malefactors and their crime. Bishop Losinga concludes the sentence of excommunication in dramatic, unique terms:

> May the flesh of those who have devoured my stag rot, as the flesh of Herod rotted, who shed the blood of Innocents in order to come to Christ; may they have their portion with Judas the traitor, with Ananias and Sapphira, with Dathan and Abiram. Let them have the Anathema Maranatha, unless they shall come to a better mind, and make me some reparation. Amen. Amen. Amen. This excommunication I pronounce, well-beloved brethren, not because a single deer is of any great importance to me, but because I am desirous that the evil-doers should repent, and come to confession, and afterwards receive meet correction for so gross a robbery.[105]

The bishop of Norwich, while clearly peeved at the death of his only deer, focuses instead on the welfare of the poachers' souls and his diocesan role as a loving, paternal disciplinarian. By taking the high ground, and putting animus before corpus, Losinga – like Becket, Bronescombe and Sutton after him – fully utilized the powers of excommunication and absolution bestowed upon him through his spiritual office, regardless of whether the poachers were ever brought to justice. Indeed, by emphasizing their power, authority and masculinity throughout their respective dioceses in such a public manner helped them to remind the laymen who wielded the sharper sword – that of anathema – in England.

In conclusion, like their martyred, archiepiscopal lord Becket, the bishops, abbots and other powerful churchmen whose private hunting grounds fell victim to violent havocking behaved manfully and donned the armor of God by outwardly dismissing the violent attacks against their parks, rising above the intended insult and punishing the invaders *not* for the sake of the deer nor the incursion – so they claimed, at least – but rather for the sake of the sinners' souls and the need for communal and spiritual reconciliation in the diocese.[106] The derisive acts committed by laymen against bishops and their animals attempted to communicate, through gendered rituals of emasculated shaming, the laymen's physical and symbolic mastery over the clerical body as well as publicize the bishop's impotence as the paterfamilias of the episcopal household and diocese. The bishops, in turn, attempted to out-perform the audacious knights with a public ceremony of their own – excommunication and defamation – which served to offset the violent message of shame that the laymen had designed, as well as to usurp the assailants' authority, manliness and power. Such meaningful acts of ritualized violence, when contextualized in their respective historical settings, provide insight into how laymen thought about themselves as a group and, conversely, how they thought about clerics in late medieval England. These rituals shed light on how laymen communicated their own masculine identity while both challenging and ridiculing clerical identity and masculinity.

Notes

1. J. C. Robertson and J. B. Sheppard, eds., *Materials for the History of Thomas Becket*, 7 vols. (London, 1875–1885), vol. 1, 55, vol. 2, 404, vol. 3, 360, vol. 4, 65, vol. 5, 152, vol. 6, 278, 300, 449, 582; Michael Staunton, *Thomas Becket and His Biographers* (Woodbridge, 2006), 154–5, 86–7, 89, 202–3.
2. Robertson and Sheppard, eds., *Materials for the History of Thomas Becket*, vol. 1, 117, vol. 2, 7, vol. 3, 24, 26, 483–5.

3. Staunton, *Thomas Becket and His Biographers*, 187–8.

4. 'Et magis confortabatur in Domino, viriliter agens et induens se armaturam Dei, ut fortiter stare posset in die Domini; sed secretum habuit ne in tanto festo tumultus fieret, quantum in eo erat' (Robertson and Sheppard, eds., *Materials for the History of Thomas Becket*, vol. 3, 130).

5. 'Iumentum in nominis mei contemptum tamquam in diminutione bestiae dehonestari possim cauda truncatum est' (Robertson and Sheppard, eds., *Materials for the History of Thomas Becket*, vol. 1, 130). Translation from George Neilson, 'Caudatus Anglicus: A Medieval Slander', *Transactions of the Glasgow Archaeological Society*, new series 2 (1896): 446. Later iconography depicts Becket upon the horse when its tail is docked (Tancred Borenius, *St Thomas Becket in Art* [London, 1970], 58–9).

6. Robertson and Sheppard, eds., *Materials for the History of Thomas Becket*, vol. 1, 61, vol. 2, 428, vol. 3, 130, vol. 4, 359, vol. 5, 83, 88, 91, 95, vol. 6, 559, 61, 73, 83–5. Becket's murderers rendezvoused at the seat of the de Broc family, Saltwood Castle, upon arriving in England from France, and Robert de Broc served as their guide to the crime (Robertson and Sheppard, eds., *Materials for the History of Thomas Becket*, vol. 1, 127–31, vol. 3, 29–30, 488).

7. Unlike the more formulaic bishops' registers of the late Middle Ages, the episcopal *memoranda* employed in this study were written in the form of an administrative diary, especially when the bishops recorded events which they deemed important or urgent – such as powerful laymen challenging their power and authority (Bronescombe, vol. 1, xiv; Rosalind M. T. Hill, 'Bishop Sutton and His Archives: A Study in the Keeping of Records in the Thirteenth Century', *Journal of Ecclesiastical History* 2 [1951]: 43).

8. Vern L. Bullough, 'On Being Male in the Middle Ages', in *Medieval Masculinities: Regarding Men in the Middle Ages*, ed. Clare A. Lees (Minneapolis, 1994), 34.

9. Jo Ann McNamara, 'The *Herrenfrage*: The Reconstruction of the Gender System, 1050–1150', in *Medieval Masculinities: Regarding Men in the Middle Ages*, ed. Clare A. Lees (Minneapolis, 1994): 3–29; Jacqueline Murray, 'Masculinizing Religious Life: Sexual Prowess, the Battle for Chastity and Monastic Identity', in *Holiness and Masculinity in the Middle Ages*, ed. P. H Cullum and Katherine J. Lewis (Cardiff, 2004): 24–42.

10. Jean Birrell, 'Deer and Deer Farming in Medieval England', *The Agricultural History Review* 40, no. 2 (1992): 115; Leonard Cantor, 'Forests, Chases, Parks and Warrens', in *The English Medieval Landscape*, ed. Leonard Cantor (Philadelphia, 1982), 77; Oliver H. Creighton, *Castles and Landscapes: Power, Community and Fortification in Medieval England* (London, 2002), 131, 90, 224; Emma Griffin, *Blood Sport: Hunting in Britain since 1066* (London, 2007), 63–6; Chester Kirby, 'The English Game Law System', *The American Historical Review* 38 (1933): 240; Susan Lasdun, *The English Park: Royal, Private and Public* (London, 1991), 8; William P. Marvin, *Hunting Law and Ritual in Medieval English Literature* (Cambridge, 2006), 108–9; William P. Marvin, 'Slaughter and Romance: Hunting Reserves in Late Medieval England', in *Medieval Crime and Social Control*, ed. Barbara A. Hanawalt and David Wallace (Minneapolis, 1999), 225–9; Colin Platt, *Medieval England: A Social History and Archaeology from the Conquest to 1600 A.D.* (London,

1994), 47; Trevor Rowley, 'Chapter Five: Woodland, Forests and Parks', in *The High Middle Ages, 1200–1550: The Making of Britain*, ed. Trevor Rowley (London, 1986), 171; Paul Stamper, 'Woods and Parks', in *The Countryside of Medieval England*, ed. Grenville Astill and Annie Grant (Oxford, 1988), 140–6.

11. The royal forest was a large tract of (wooded) country belonging to the king, and subject to forest law. The primary purpose of the royal forest was to protect animals for the king to hunt (Richard W. Kaeuper, 'Forest Law', in *A Dictionary of the Middle Ages*, ed. Joseph Strayer [New York, 1985], vol. 5, 127–31; Charles R. Young, *The Royal Forests of Medieval England* [Leicester, 1979], 21–2, 36–7, 65–6). The chase was a private forest that a select few of the great lords were allowed to maintain on their lands. The chase, unlike the royal forest, was subject to common law instead of forest law. Although the terms chase and forest are sometimes used synonymously, there were legal distinctions, and no one was allowed to hunt on a chase without the lord's permission. The warren was a game park stocked with various smaller animals, especially hares and rabbits. (Cantor, 'Forests, Chases, Parks and Warrens', 70, 82; Raymond Grant, *The Royal Forests of England* [Gloucester, 1991], 29–32.)

12. Leonard Cantor, 'The Medieval Hunting Grounds of Rutland', *Rutland Record* 1 (1980): 14; Rowley, 'Woodland, Forests and Parks', 182–3.

13. Cantor, 'Forests, Chases, Parks and Warrens', 76–7; Roger B. Manning, *Hunters and Poachers: A Social and Cultural History of Unlawful Hunting in England, 1485–1640* (Oxford, 1993), 176; Platt, *Medieval England: A Social History and Archaeology*, 47. For specific examples of a lord's parks and their maintenance, see P. Franklin, 'Thornbury Woodlands and Deer Parks: The Earl of Gloucester's Deer Parks', *Transactions of the Bristol and Gloucestershire Archaeological Society* 107 (1982): 149–69; Edward Roberts, 'The Bishop of Winchester's Deer Parks in Hampshire, 1200–1400', *Proceedings of the Hampshire Field Club Archaeological Society* 44 (1988): 67–86.

14. Jean Birrell, 'Procuring, Preparing, and Serving Venison in Late Medieval England', in *Food in Medieval England: Diet and Nutrition*, ed. C. M. Woolgar, D. Serjeantson, and T. Waldron (Oxford, 2006), 177; Leonard Cantor, 'The Medieval Parks of Leicestershire', *Transactions of the Leicestershire Archaeological and Historical Society* 46 (1970–71): 9; Creighton, *Castles and Landscapes*, 131, 90, 224; Platt, *Medieval England: A Social History and Archaeology*, 47. See also Leonard Cantor and J. Hatherly, 'The Medieval Parks of England', *Geography* 64 (1979): 71–85.

15. Birrell, 'Deer and Deer Farming in Medieval England', 114–16; Creighton, *Castles and Landscapes*, 190; Griffin, *Blood Sport*, 63–6; John Hatcher, *Rural Economy and Society in the Duchy of Cornwall, 1300–1500* (Cambridge, 1970), 180; Marvin, 'Slaughter and Romance', 243; Stamper, 'Woods and Parks', 140–6.

16. Manning, *Hunters and Poachers*, 176.

17. Record Commission, ed., *Statutes of the Realm* (London, 1810–1828), vol. 1, 111–12.

18. Cantor, 'Forests, Chases, Parks and Warrens, 78–9.

19. Other examples include the future Edward II (1307–1327) who, while still a
    prince and traveling in the company of his companion Piers de Gaveston,
    infamously invaded a park belonging to Bishop Langton, King Edward I's
    treasurer; they pulled down the park's palings and wreaked havoc upon the
    bishop's deer and other animals. Langton, a trusted advisor of the boy's
    father, dolefully complained to his lord about the insult. The king was
    furious; he ostracized the prince for months (Thomas Costain, *The Three
    Edwards* [New York, 1958], 105–6; Hilda Johnstone, *Edward of Carnarvon,
    1284–1307* [Manchester, 1946], 96–102). During the reign of King Edward
    III (1327–1377), some of the chancellor's men broke a park belonging to the
    bishop of Worcester and killed a number of his deer and swans (CPR 1350–
    1354, 79). Similarly, during the reign of Richard II (1377–1399), a knight,
    his son and a number of supporters repeatedly attacked a park belonging
    to the bishop of Durham; they slaughtered all the deer they could find,
    destroyed crops belonging to the tenants and then 'commanded them by
    their messages to the parkers of the park that if they should be so bold as to
    come to the park or to pass the town of Crake to defeat them of their hunt,
    they should be killed without mercy' (William Baildon, ed., *Select Cases
    in Chancery, 1364–1471* [London, 1896], 18–19.). Even bishops resorted to
    such violence during disputes; in the mid-thirteenth century the bishop of
    Durham broke down a park belonging to the prior and convent of Durham,
    killed deer and drove off the rest (Evelyn P. Shirley, *Some Account of English
    Deer Parks* [London, 1867], 227).
20. E.g. CPR 1348–1350, 174, CPR 1374–1377, 156, CPR 1391–1396, 235, CPR
    1422–1429, 124.
21. L. F. Casson, ed., *The Romance of Sir Degrevant: A Parallel-Text Edition from
    Mss. Lincoln Cathedral A. 5. 2 and Cambridge University Ff. I. 6, vol. 221, Early
    English Text Society* (Oxford, 1970).
22. William P. Marvin has analyzed the fictitious earl's destruction of Sir Degre-
    vant's parks, arguing in part that the earl wished to create a crisis in which
    Sir Degrevant would face great shame and impotence through the lack of
    response to such violence; Degrevant was on crusade and subsequently
    forced to return to defend his parks and honor (Marvin, 'Slaughter and
    Romance', 226–9, 39).
23. N. J. Sykes, 'The Impact of the Normans on Hunting Practices in England',
    in *Food in Medieval England: Diet and Nutrition*, ed. C. M. Woolgar, D. Ser-
    jeantson, and T. Waldron (Oxford, 2006), 164; C. M. Woolgar, *The Great
    Household in Late Medieval England* (New Haven, 1999), 196.
24. Griffin, *Blood Sport*, 29–31; Anne Rooney, *Hunting in Middle English Literature*
    (Cambridge, 1993), 2–5; John Steane, 'Chapter Six: Formalized Violence:
    Hunting, Hawking and Jousting', in *Archaeology of the Medieval English
    Monarchy* (London, 1999), 146. See John Cummins, *The Hound and the
    Hawk: The Art of Medieval Hunting* (New York, 1988); Henry L. Savage, 'Hunt-
    ing in the Middle Ages', *Speculum* 8, no. 1 (1933): 30–41; M. Thièbaux, *The
    Stag of Love: The Chase in Medieval Literature* (Ithaca, 1974).
25. Nelson emphasizes that this story is about choosing between a life secluded
    from the world in a monastery and one active within it – like any other
    noble layman (J. L. Nelson, 'Monks, Secular Men and Masculinity, c. 900',
    in *Masculinity in Medieval Europe*, ed. D. M. Hadley [London, 1999], 133).

26. Manning, *Hunters and Poachers*, 176. Even the University of St Andrews, 'with unwonted liberality actually allowed its students [clerics] to go a-hawking, provided they went in their own clothes and not in 'dissolute habiliments borrowed from lay cavaliers' (Hastings Rashdall, *The Universities of Europe in the Middle Ages*, 2 vols. [Oxford, 1895], vol. 2, part 2, 675. See also W. Brown, ed., Register of William Wickwane, *Archbishop of York, 1279–1285*, vol. 114 [Surtees Society, 1907], 88; Geoffrey Chaucer, *The Riverside Chaucer*, ed. Larry D. Benson, 3 ed. [Boston, 1987], General Prologue, ll. 189–92; H. O. Cox, ed., *Poema Quod Dicitur Vox Clamantis, Necnon Chronica Tripartita* [London, 1850], vol. 3, ll. 1490–512; F. D. Matthew, ed., *The English Works of Wyclif, Hitherto Unprinted* [New York, 1975], 212–13, 49; G. R. Owst, *Preaching in Medieval England: An Introduction to Sermon Manuscripts of the Period, c. 1350–1450* [Cambridge, 1926], 279).

27. In one study, about 5 per cent of poachers were clerics (J. Charles Cox, *The Royal Forests of England* (London, 1905), 155–6. Examples of clerical poachers include: W. W. Capes, *Scenes of Rural Life in Hampshire among the Manors of Bramshott* (London, 1901), 48–50; F. C. Hingeston-Randolph, ed., *The Register of Thomas Brantyngham, Bishop of Exeter (A.D. 1370–1394)*, 2 vols. (London, 1901–1906), vol. 2, 637; A. F. Leach, ed., *Visitations and Memorials of Southwell Minster*, vol. 48 (Camden Society, ns: 1891), 62; Shirley, *Some Account of English Deer Parks*, 227.

28. Marvin, 'Slaughter and Romance', 226–9.

29. Cantor, 'Forests, Chases, Parks and Warrens', 78–9; G. J. Turner, ed., *Select Pleas of the Forest*, vol. 13 (Selden Society, 1901), xli.

30. David Crook, 'The 'Petition of the Barons' and Charters of Free Warren, 1227–1258', in *Thirteenth Century England Viii: Proceedings of the Durham Conference, 1999*, ed. Michael Prestwich, R. H. Britnell, and Robin Frame (Boydell, 2001), 46.

31. Diana E. Greenway and Jane E. Sayers, eds., *Jocelin De Brakelond: Chronicle of the Abbey of Bury St Edmunds* (Oxford, 1989), 26.

32. Edmund has received relatively little attention from scholars. He was the first son of King Henry III's brother, Richard, Earl of Cornwall, by his second wife Sancha (the sister of Henry III's wife, Eleanor, and daughter of Raymond Berengar, count of Provence (Laura M. Midgley, 'Edmund Earl of Cornwall and His Place in History' [M.A., University of Manchester, 1930]).

33. *CChR 1257–1300*, 16.

34. Leonard Elliott-Binns, *Medieval Cornwall* (London, 1955), 246, 99–300; Charles Henderson, *Essays in Cornish History* (Oxford, 1935), 157–9; James Whetter, *Cornwall in the Thirteenth Century: A Study in Social and Economic History* (Lostwithiel, 1998), 159–62. At least until the last century, there was an area in the parish of St Allen called 'Nancarrow' or 'Deer Valley' (Davies Gilbert, *The Parochial History of Cornwall*, 4 vols. [London, 1838], vol. 1, 15–21; Arthur Grigg, *Place-Names in Devon and Cornwall* [Plymouth, 1988], 13–14).

35. John Cummins, 'Veneurs S'en Vont En Paradis: Medieval Hunting and the 'Natural' Landscape', in *Inventing Medieval Landscapes: Senses of Place in Western Europe*, ed. John Howe and Michael Wolfe (Gainesville, FL, 2002), 47.

36. Creighton, *Castles and Landscapes*, 131, 224.

37. Glasney, a prominent collegiate house in Cornwall, was prominently situated at the foot of the bishop's deer park at Penryn; Bronescombe provided great endowments in order to make it a diocesan stronghold (Anthony Emery, *Greater Medieval Houses of England and Wales, 1300–1500* [Cambridge, 2006], 450, n. 39; F. E. Halliday, *A History of Cornwall*, 2 ed. [London, 1975], 137).
38. Chroniclers such as Trivet claimed that stag-hunting was the king's favorite sport; it seems certain from the documents, however, that falconry was Edward I's true love (Michael Prestwich, *Edward I*, 2nd ed. [New Haven, 1997], 115).
39. According to the Caption of Seisin of 1337 Launceston held 15 deer, Trematon 42, Carrybullock 150, Helsbury and Lantecos 180, Liskeard 200, Restormel 300 and an unknown number at Penlyne (Elliott-Binns, *Medieval Cornwall*, 164; Hatcher, *Rural Economy and Society*, 179; P. L. Hull, *The Caption of Seisin of the Duchy of Cornwall, 1337* [Exeter, 1971]).
40. Edmund enlarged his park at Restormel in 1297/98 by 3.5 Cornish furlongs, at the expense of 32s. 6d. in rents. The earl had also enlarged his park at Penlyne in 1296/97, causing 'decay' in the rents of the assize of 6s. 4.5d. (Hatcher, *Rural Economy and Society*, 179 n. 3 and 4).
41. Bronescombe, vol. 1, xliii.
42. The first violent attack against the bishop's interests by members of the earl's household took place against Hartland parish church sometime before 23 July 1272, during which blood was spilt in the church (Bronescombe, vol. 2, 54, #894; *CPR 1272–1281*, 123).
43. The following spring Bishop Bronescombe departed to meet King Edward I in Gascony (the king was returning from Crusade to begin his reign). Bronescombe was overseas between May 1273 and August/September 1274 (*CPR 1272–1281*, 9, 30, 46). About 1 month after Bronescombe had departed, Edmund, Earl of Cornwall, also departed to meet his lord in Gascony (*CPR 1272–1281*, 11).
44. Because of the remote nature of Cornwall, the region's sheriffs were exceptionally influential; while the earl was away from the county, the sheriff of Cornwall had personal custody of deer parks and escheats (William A. Morris, *The Medieval English Sheriff to 1300* [Manchester, 1968], 93–4, 181–2; Robert C. Palmer, *The County Courts of Medieval England, 1150–1350* [Princeton, 1982], 28–55).
45. CPR 1272–1281, 120; Bronescombe, vol. 2, 57, #924; 72, #1024; 76–7, #1058; 77–9, #1067; 79–80, #1069; 80–1, #1079; 81, #1082; 81, #1083; 81, #1084; 82–3, #1085; 83–4, #1086; 87–8, #1100; 96–7, #1156; 137–8, #1429. Bishop Bronescombe's successors also had to defend their parks against similar attacks. In 1311, for example, the episcopal parks of Pawton and Lanner and the free-warrens at Pawton, Penryn, Tregear, St Germans and Lawhitton were attacked and the bishop's servants assaulted (Elliott-Binns, *Medieval Cornwall*, 300). See also F. C. Hingeston-Randolph, ed., *The Register of John De Grandisson, Bishop of Exeter (1327–69)*, 3 vols. (London, 1899), vol. 1, 352–3, 466, 558–9; Hingeston-Randolph, ed., *The Register of Thomas Brantyngham*, vol. 1, 311 and vol. 2, 650.

46. *CPR 1272–1281, 120*; Bronescombe, vol. 2, 57, #924, 76, #1058, 81, #1082, #1083.

47. Bronescombe, vol. 2, 82–83, #1085, vol. 2, 96–7, #1156. The bishop absolved those who had been dragooned into attacking his parks and clerics (Bronescombe, vol. 2, 79–80, #1069).

48. Elliott-Binns, *Medieval Cornwall*, 303; Emery, *Greater Medieval Houses*, 550–51; Hull, *The Caption of Seisin of the Duchy of Cornwall, 1337*, xxii–xxiii.

49. By 1289, John St John's household was continually staying with the king (*CCR 1288–1296*, 25). See also: Charles Moor, *Knights of Edward I*, 5 vols. (Harleian Society, 1929–1932), vol. 4, 177–9.

50. *CPR 1272–1281, 211*; Richard Huscroft, ed., *The Royal Charter Witness Lists of Edward I (1272–1307)*, vol. 279, List and Index Society (Wiltshire, 2000), 71–94.

51. *LRS 52*, 104–5, 107–9, 117–18; Norma Adams and Charles Donahue, eds., *Select Cases from the Ecclesiastical Courts of the Province of Canterbury: C. 1200–1301*, vol. 95, *Selden Society* (London, 1981), 587–98; John Le Neve, *Fasti Ecclesiae Anglicanae, 1066–1300*, 3 vols., vol. 3: Lincoln (London, 1977), 65, 102; Public Record Office, *Calendar of Inquisitions Miscellaneous (Chancery), A.D. 1219–1307* (Kraus Reprint, 1973), #1640, 459–60; G. O. Sayles and H. G. Richardson, eds., *Rotuli Parliamentorum Anglie Hactenus Inediti 1279–1373*, Camden Third Series, vol. 51 (London, 1935), 101–02.

52. Moor, *Knights of Edward I*, vol. 4, 161–2. Paynel became the general attorney and keeper of Sir John St John's lands in 1297 and, when Sir John died in 1302, his estate provided hundreds of pounds to Sir Thomas for services rendered (*CPR 1292–1301*, 294–95; *CCR 1296–1302*, 608).

53. Gosberton church was worth over 33 pounds annually, with a pension of over 26 pounds owed to the chapter of Lincoln Cathedral (http://www.hrionline.ac.uk/db/taxatio/printbc.jsp?benkey=LI.LK.HD.20).

54. A. R. Maddison, ed., *Lincolnshire Pedigrees*, 4 vols. (Harleian Society, 1902–1906), vol. 3, 838–9; W. P. W. Phillimore and Francis N. Davis, eds., *Rotuli Hugonis De Welles, Episcopi Lincolniensis, A. D. Mccix–Mccxxxv*, 3 vols., *Lincoln Record Society: Vols. 3, 6, 9* (Lincoln, 1912), vol. 3, 172, 75. A badly disfigured effigy of Sir Ranulph's father, John, still lies in the church's south transept (Walter J. Kaye, *A Brief History of the Church and Parish of Gosberton in the County of Lincoln* [London, 1897], 105–06).

55. Ranulph also promised to pay to the cathedral an annual pension of 40 marks from the church of Gosberton (*LRS 39*, 202).

56. *CPR 1292–1301*, 170, 212. According to the bishop's register: '…omnia arma et quecumque alia quibus sive pro municione corporum sive constructione machinarum sive alio modo in intrusione dicte ecclesie usi fuerant…' (*LRS 60*, 122–3). Apparently, using siege equipment was an effective means of attacking or defending churches in the diocese of Lincoln in the 1290s; a similar event occurred at the church of Claybrooke in 1295 (*LRS 60*, 112–14).

57. Thomas Paynel (*CPR 1292–1301*, 250, 290; *CCR 1288–1296*, 96); John St John (*CCR 1288–1296*, 18, 176, *CPR 1281–1292*, 510; Moor, *Knights of Edward I*, vol. 4, 175–7).

58. Interestingly, both Edmund, Earl of Cornwall and Bishop Sutton of Lincoln helped witness these privileges (*CChR 1257–1300*, 255; *CFR 1272–1307*, 122; Charles Roberts, ed., *Calendarium Genealogicum. Henry Iii and Edward I*, 2 vols. [London, 1865], vol. 1, 296). Sir Ranulph served as sheriff of Lincoln between 1309 and 1311; during his tenure he continued to harass numerous powerful churchmen – particularly abbots – and in one case seized over 100 livestock, starving and killing many of the beasts (*CPR 1307–1313*, 474, 475, 529). Contemporary political songs lament similar behavior by medieval English sheriffs (Thomas Wright and Peter R. Coss, eds., *Thomas Wright's Political Songs of England: From the Reign of John to That of Edward II* [Cambridge, 1996], 229).

59. *CCR 1288–1296*, 473–4. Intriguingly, at almost exactly the same time Bishop Sutton's park at Lyddington, which is relatively near Sherwood Forest, was hit by (unidentified) poachers (*LRS 60*, 136). It is entirely possible that these poachers were Sir Ranulph Rye and his comrades.

60. '…prioris et conventus Spalding quod novus locus in parochia de Pyncebeck' vulgariter nuncupatru…' (*LRS 52*, 78–9).

61. Land became even more valuable in the late thirteenth century because there was a great deal of inundation from the sea along the coast (William Page, ed., *A History of the County of Lincoln, Volume 2* [The Victoria History of the Counties of England, 1906], 118–24).

62. *CChR, 1226–1257*, 217; H. E. Hallam, ed., *The Agrarian History of England and Wales: Volume 2 (1042–1350)* (Cambridge, 1988), 538; H. C. Maxwell Lyte, ed., *Liber Feodorum. The Book of Fees, Commonly Called Testa De Nevill*, 3 vols. (Kraus Reprint, 1971), vol. 2, 1008.

63. Page, ed., *A History of the County of Lincoln, Volume 2*, 118–24.

64. *LRS 52*, 70–2.

65. '…usque ad sexaginta et amplius tum per canes tum per alia ingenia non sine excogitata malicia occiderunt, et taliter occisos pro majori parte quo eis libuit temere asportarunt, in contumeliam hostie salutaris et vive, diabolo ad irreverenciam ipsius temporis sacri nephandum piaculi sacrificium in sanguine pecorum immolantes de quo vehemens scandalum in dictis partibus dicitur suscitatum…' (*LRS 52*, 78–9).

66. The editors of Sutton's and Bronescombe's episcopal registers make no connection between the invasions of the bishops' parks and the contentious struggles over Cornish rights and liberties and the prebend of Thame, even though the participants knew one another and shared similar tactics, violence and motives (*LRS 52*, 78–79 n. 2; Bronescombe, vol. 1, xliv).

67. Barbara A. Hanawalt, 'Men's Games, King's Deer: Poaching in Medieval England', in *'of Good and Ill Repute': Gender and Social Control in Medieval England*, ed. Barbara A. Hanawalt (Oxford, 1998), 150–1; Jennifer Thibodeaux, 'Odo Rigaldus, the Norman Elite, and the Conflict over Masculine Prerogatives in the Diocese of Rouen', *Essays in Medieval Studies* 23 (2006): 41–55. Important studies on poaching and forest related disputes include: Jean Birrell, 'Aristocratic Poachers in the Forest of Dean: Their Methods, Their Quarry and Their Companions', *Transactions of the Bristol and Gloucestershire Archaeological Society* 119 (2001): 147–54; Jean Birrell, 'Common Rights in the Medieval Forest: Disputes and Conflicts in the Thirteenth Century', *Past and Present* 117 (Nov. 1987): 22–49; Jean Birrell, 'Peasant

Deer Poachers in the Medieval Forest', in *Progress and Problems in Medieval England: Essays in Honour of Edward Miller*, ed. Richard Britnell and John Hatcher (Cambridge, 1996), 68–88; F. H. M. Parker, 'Some Stories of Deer Stealers', *Transactions of the Cumberland and Westmoreland Antiquarian and Archaeological Society*, N. S. 7 (1907): 1–30; Derek Rivard, 'The Poachers of Pickering Forest, 1282–1338', *Medieval Prosopography* 27 (1996): 97–144.

68. Jean Birrell, 'Who Poached the King's Deer? A Study in Thirteenth-Century Crime', *Midland History* 7 (1982): 15–16; Hanawalt, 'Men's Games, King's Deer', 151–5; Marvin, 'Slaughter and Romance', 230–6.

69. The violence that took place in Cornwall in the 1270s was also havocking, not poaching; the Earl of Cornwall's household mustered a large force and broke down the enclosures of the bishop of Exeter's parks, killed deer and chased away others. As part of their settlement and reconciliation, Edmund, Earl of Cornwall, conceded that the enclosures surrounding Bishop Bronescombe's deer parks should be repaired and restocked (Bronescombe, vol. 2, 77–9, #1067, 96–7, #1156).

70. Queen Elizabeth I famously havocked 27 deer belonging to Lord Berkeley after he slighted her during a royal progress in 1572 (Manning, *Hunters and Poachers*, 47–9). In medieval England, large bands of poachers also took advantage of a prominent lord's death to hunt in the deceased man's forests and poach explicitly to protest politically against forest laws or rights of free-warren. Such attitudes can be seen through local reaction to the death of Henry III (1272) in the West Midlands, for example, when several local men entered the local forest with greyhounds, bows and arrows and proceeded to spend the day hunting and poaching with great furor (Birrell, 'Who Poached the King's Deer?', 15; Hanawalt, 'Men's Games, King's Deer', 153). Likewise, shortly after Edmund, Earl of Cornwall died (1300), a number of people entered his deer parks at Oakham and Cold Overton in the diocese of Lincoln – while they were in King Edward I's hand – and hunted and carried away deer (*CPR 1292–1301*, 626, 630).

71. Birrell, 'Who Poached the King's Deer?', 18–19.

72. Rivard, 'The Poachers of Pickering Forest, 1282–1338', 97–8.

73. When royal or archiepiscopal records are available to corroborate what Bishop Sutton's register states on a variety of different matters, they are largely in accord; therefore, there is no reason to doubt the bishop when he records that upward of five dozen deer were slaughtered at Pinchbeck.

74. 'per canes tum per alia ingenia.' Both hunters and poachers on occasion drove deer into nets and this method was not considered an improper way of taking deer (Birrell, 'Who Poached the King's Deer?', 17–18; Jean Birrell, ed., *Records of Feckenham Forest, Worcestershire, C. 1236–1377* [Worcestershire Historical Society, 2006], xix, 71, 108, 55, 65–6, 69, 70).

75. Birrell, 'Who Poached the King's Deer?', 17–18; Griffin, *Blood Sport*, 30–1; Hanawalt, 'Men's Games, King's Deer', 147; Savage, 'Hunting in the Middle Ages', 36–8.

76. '…et taliter occisos pro majori parte quo eis libuit temere asportarunt…' (*LRS 52*, 78–9).

77. Birrell, 'Deer and Deer Farming in Medieval England', 126; Birrell, 'Who Poached the King's Deer?', 11, 19; Steane, 'Formalized Violence: Hunting, Hawking and Jousting', 148.

78. Woolgar, *The Great Household in Late Medieval England*, 115. Gifts of venison by noblemen to ecclesiastical bodies on religious holidays could be grand affairs indeed; for example, the knightly family of Le Baud had a tradition dating back to the late thirteenth century in which a buck was carried by way of procession to the high altar of St Paul's in London; the tradition continued into the sixteenth century (Mackenzie E. C. Walcott, *Traditions and Customs of Cathedrals* [London, 1872], 131–2).

79. 'contra voluntatem dominorum seu custodum loci ejusdem noctanter ingressi' (*LRS 52*, 78–9). Recall that Birrell, Hanawalt, and Marvin have analyzed the connection between emasculating a lord's parkers and shaming the lord himself (Birrell, 'Who Poached the King's Deer?', 15–16; Hanawalt, 'Men's Games, King's Deer', 151–5; Marvin, 'Slaughter and Romance', 230–6).

80. '...in contumeliam hostie salutaris et vive, diabolo ad irreverenciam ipsius temporis sacri nephandum piaculi sacrificium in sanguine pecorum immolantes de quo vehemens scandalum in dictis partibus dicitur suscitatum...' (*LRS 52*, 78–9).

81. R. N. Swanson, 'Angels Incarnate: Clergy and Masculinity from Gregorian Reform to Reformation', in *Masculinity in Medieval Europe*, ed. D. M. Hadley (London, 1999), 160.

82. Andrew G. Miller, 'Violence and Vestments: 'Defrocking' and 'Refrocking' Clerics in Medieval England' (under review); Andrew G. Miller, 'A Horse's Ass: Mutilating Tails and the Message of Masculinity in the Middle Ages' (under review).

83. Consider a classic Icelandic example: Egil, during a dispute with King Eirik, Egil cursed the king and his kingdom and, upon parting his realm stuck a horse's head upon a hazel staff on a rocky promontory overlooking the kingdom. He then recited a prayer, facing the head toward the land, and carved runes into the staff (Snorri Sturluson, *Egil's Saga*, ed. Hermann Palsson, Paul Edwards and Paul G. Edwards [Penguin Classics, 1976], 148.) Likewise, the Danish chronicler Saxo Grammaticus relates how disputants stuck a horse's head on a pole, with its 'jaws grinning agape', in order to unnerve their opponents. One of the parties, upon witnessing the grave insult, and 'understanding the whole foul contrivance...bade his men keep silent and behave warily' (Frederick Y. Powell and Rasmus B. Anderson, eds., *The Nine Books of the Danish History of Saxo Grammaticus*, 2 vols., vol. 1 [London, 1905], 281). Also see Jeffrey Jerome Cohen, 'Decapitation and Coming of Age: Constructing Masculinity and the Monstrous', in *The Arthurian Yearbook*, ed. Keith Busby (New York, 1993), 173–92.

84. Edward Berry, *Shakespeare and the Hunt: A Cultural and Social Study* (Cambridge, 2001), 73–5; Griffin, *Blood Sport*, 31. During the English Civil War, a common method for showing animosity against powerful men of the opposing side was by invading their parks and killing their deer; a certain George Melsam and his brothers claimed to have cut off the heads of 300 deer, which were found at the bottom of a well (W. H. P. Greswell, *The Forests and Deer Parks of the County of Somerset* [Taunton, 1905], 11).

85. Rivard, 'The Poachers of Pickering Forest, 1282–1338', 97–8.

86. Berry, *Shakespeare and the Hunt: A Cultural and Social Study*, 19; Manning, *Hunters and Poachers*, 41.

87. Birrell, 'Who Poached the King's Deer?', 15–16; Hanawalt, 'Men's Games, King's Deer', 153; Turner, ed., *Select Pleas of the Forest*, 38–9.

88. Hanawalt, 'Men's Games, King's Deer', 153.

89. My emphasis. The invaders also cut down trees and took them and other goods belonging to the archbishop (*CPR 1391–1396*, 235, 238).

90. Birrell, 'Deer and Deer Farming in Medieval England', 116; Griffin, *Blood Sport*, 56; Marvin, *Hunting Law and Ritual in Medieval English Literature*, 78.

91. John Aberth, *From the Brink of the Apocalypse: Confronting Famine, War, Plague, and Death in the Later Middle Ages* (London, 2001), 147–8.

92. John E. Archer, *By a Flash and a Scare: Arson, Animal Maiming, and Poaching in East Anglia, 1815–1870* (Oxford, 1990), 199; Timothy Shakesheff, *Rural Conflict, Crime and Protest: Herefordhsire, 1800–1860* (Woodbridge, 2003), 176–200, esp. 198.

93. Joyce E. Salisbury, *The Beast Within: Animals in the Middle Ages* (New York, 1994), 40–1.

94. Stephen D. White, 'The Politics of Anger', in *Anger's Past: The Social Uses of an Emotion in the Middle Ages*, ed. Barbara Rosenwein (Ithaca, 1998), 139.

95. Carol Symes, *A Common Stage: Theater and Public Life in Medieval Arras* (Cornell, 2007), 137–8. See esp. 127–82. I would like to thank Sharon Farmer for calling my attention to this study.

96. Symes, *A Common Stage: Theater and Public Life in Medieval Arras*, 149–51.

97. Bronescombe, vol. 2, 72, #1024; 74, #1044; 81, #1084; 88–89, #1106; 93, #1130; 137–8, #1429 (Latin in vol. 3, 46–8, #1429); castration mentioned in vol. 2, 76–7, #1058). See also Bronescombe, vol. 2, 80–1, #1079; 81, #1082; 81, #1083; 83–4, #1086; 96–7, #1156. As the knights' pledges demonstrate, in canon law two conflicting households would bind themselves to arbitrators and agree upon sums of money to indicate their willingness to abide by the agreed-upon arbitration (Edward Powell, 'Arbitration and the Law in England in the Late Middle Ages', *Transactions of the Royal Historical Society, fifth series* 33 [1983]: 53–5).

98. Bronescombe, vol. 2, 82–3, #1085; 96–7, #1156.

99. '*Sentencia pro Interfectione Damorum Prioris Spalding*' (*LRS 52*, 78–9).

100. *LRS 60*, 122–3; Rosalind M. T. Hill, *Oliver Sutton, Dean of Lincoln, Later Bishop of Lincoln (1280–99)* (Lincoln, 1950), 23.

101. I assume that Sir Thomas Paynel was likewise upset with such demands and penalties and had secured some sort of settlement with the bishop, as did Sir Ranulph. Because Sir Thomas appears to have been a knight of greater status than Sir Ranulph, it is possible that Sir Thomas paid some sort of fine to the bishop and had all (public) penance dropped entirely.

102. *LRS 60*, 168–70.

103. *LRS 60*, 201–2; Rosalind M. T. Hill, 'Public Penance: Some Problems of a Thirteenth-Century Bishop', *History*, new series 36 (1951): 221–2; Robert E. Rodes, *Ecclesiastical Administration in Medieval England: The Anglo-Saxons to the Reformation* (Notre Dame, 1977), 92–3.

104. Other examples include: Bishop Cantilupe of Hereford (1275–1282) excommunicated a trio of foresters of the Earl of Gloucester who had 'maliciously impeded' (*maliciose impediverunt*) the bishop from taking into his possession the chase of Malvern (R. G. Griffiths, ed., *The Register of Thomas De Cantilupe, Bishop of Hereford (A.D. 1275–1282)* Hereford, 1906], 227–8).

Likewise, in 1347 Sir Robert Morley killed deer belonging to the bishop of Norwich, who excommunicated him and assigned him a penance of walking barefoot and carrying a large wax taper through the streets of Norwich (Douglas Richardson, *Plantagenet Ancestry: A Study in Colonial and Medieval Families* [Baltimore, 2004], 134).

105. Edward M. Goulburn and Henry Symonds, eds., *The Life, Letters, and Sermons of Bishop Herbert De Losinga (B. Circ. A.D. 1050, D. 1119)*, 2 vols. (Oxford, 1878), vol. 1, 170–3.
106. Mary C. Mansfield, *The Humiliation of Sinners: Public Penance in Thirteenth-Century France* (Ithaca, 1995), 266–77.

10
Mirror of the Scholarly (Masculine) Soul: Scholastics, Beguines and Gendered Spirituality in Medieval Paris

Tanya Stabler Miller

The medieval university was unquestionably a man's world. In her recent study of masculinity at the medieval university, Ruth Karras has argued that the absence of women at the university was 'the most salient feature with regard to masculine identity'.[1] Studies of the medieval university, which rarely mention women, seem to support this view. The university appears as a world without women where scholars asserted their masculinity not by impressing and dominating women, but by demonstrating their superior intellectual abilities over their peers in the disputation.[2] Competition with other clerics and a sense of superiority over the rest of society were particularly intense among theologians at the University of Paris. As Ian Wei has argued, by the thirteenth century, masters of theology at the University of Paris developed a distinct group identity emphasizing their theological expertise, placing themselves at the apex of an intellectual hierarchy and claiming an authority that to some degree challenged that of major prelates.[3] Similarly, Sharon Farmer has shown that theologians constructed a 'hierarchy of masculinities', arguing for the superiority of intellectual labor over the manual labor performed by uneducated men.[4]

These expressions of clerical masculinity articulated in scholarly treatises, however, seemed to clash with the clergy's responsibilities to care for the laity. The responsibility to instruct the laity in correct doctrine was carried out in the priestly offices of preaching and hearing confession, tasks over which clerics claimed an exclusive monopoly.[5] Yet there existed a fundamental tension between pastoral care and intellectual pursuits, which became particularly problematic for members of the secular clergy in the thirteenth century as they faced fierce competition from the mendicant orders both in the realm of scholarly expertise and relations with the laity.[6]

As a 'man of the Church', the ideal university master was expected to exhibit morality and humility. As a scholar competing for higher offices and a limited pool of benefices, however, he also needed to promote his own reputation for intellectual mastery.[7] This intellectual mastery, which was expressed primarily in quodlibetal disputations, encouraged competition and fostered pride, obvious threats to the scholar's immortal soul.[8] Moreover, preoccupation with scholastic learning and competition with other men undermined the secular clergy's ability to provide effective pastoral care to the laity.[9] As preachers and confessors in contact with the laity, secular clerics were expected to convey a 'holy simplicity' as representatives of God and earthly authorities.[10] While members of the mendicant orders conveyed this simplicity through their vows of poverty, saintly founders, obedience to the papacy and relationships with inspired women, the secular clergy faced particular challenges negotiating these conflicting ideals.[11]

The secular clergy, ordained priests supported by the tithes collected from the parishes they served, were expected to provide the laity with pastoral care while maintaining their image as moral and knowledgeable pastors. They were also expected to look and behave in a manner distinct from laymen in that they were prohibited from engaging in sexual activity, frequenting taverns or hunting. In other words, secular clerics were expected to forgo activities characteristic of lay masculinity, while remaining in contact with laypeople. While the secular clergy did not always fulfill these expectations, they developed a distinct clerical masculinity based on notions of self-restraint and heroic sacrifice.[12] Yet, the gender identity of the secular cleric was fraught with contradiction, particularly on the university level, where clerics negotiated the opposing obligations and expectations associated with their scholarly activities and pastoral duties. Moreover, at all levels of the secular clergy, pastoral responsibilities brought its members into contact with women. Such contact threatened the secular clerics' vows of celibacy as well as their reputations with the laity.

Nevertheless, clerics were expected to provide pastoral care to religious women and some even felt that they benefitted from their relationships with such women, both on a personal level and in the carrying out of their professional responsibilities.[13] Certainly, the university itself excluded women and, as Karras argues, women were rarely mentioned in scholastic disputations.[14] Women, both real and abstract, however, featured prominently in what we might call the masculinity of the pastoral theologian. Indeed, it was precisely those aspects of clerical learning that excluded women, such as the disputation, that came under the criticism of clerics concerned about the scholarly preoccupation with abstract learning.

This essay will examine the ways in which one of the most influential secular clerics of the thirteenth century, Robert of Sorbon, negotiated these conflicting ideals of clerical masculinity by utilizing ideas about religious laywomen, specifically beguines.[15] Beguines were laywomen who took personal, informal vows of chastity and pursued a life of contemplative prayer combined with active service in the world. The beguines may seem an unusual model for the secular clergy. Although popularly regarded as religious women, the beguines attracted negative attention from ecclesiastical officials uncomfortable with the temporary nature of the beguines' vows, the community's lack of strict enclosure, and the absence of a papally approved rule.[16] The visibility of the beguines in an urban environment like Paris made them frequent targets among skeptics of the women's religious sincerity and commitment to celibacy.[17] The secular master William of Saint-Amour, for example, famously attacked beguines in his treatises denouncing mendicant privileges during the secular-mendicant conflict of the 1250s.[18] Nevertheless, Robert of Sorbon believed that beguines could serve as models for the masculine, pastoral spirituality he hoped to promote among the secular clergy.

Thus, despite the absence of women at the university, it is clear that university clerics needed women – or at least ideas about women – to negotiate the ideals and expectations associated with their status as male clerics. As John Coakley has demonstrated, some clerics established close friendships with religious women. Such relationships worked both to assert clerical masculinity by contrasting the cleric's institutional authority with the female penitent's divinely granted, informal authority and in turn allowed the cleric to benefit from the woman's direct relationship to God. These relationships could also, however, threaten clerical authority by raising questions about the clerics' vows of chastity. Conscious of this danger, some religious men found it useful to employ ideas about women and femininity, rather than associate with actual women, in their efforts to express their love for God or anxieties about their authoritative positions.[19]

Robert of Sorbon employed ideas about beguines in order to soften the aggressive masculinity he witnessed in the schools, which he viewed as detrimental to the clergy's relationship with the laity and – more importantly – to God. Yet, there is no evidence that Robert cultivated personal relationships with individual women, nor did he deliberately adopt explicitly feminine imagery in his treatises and sermons calling for moral reform. Instead, Robert took care to protect the male clergy's pastoral authority, which relied on a separation from actual women and assertion of masculine identity, by embracing the male form of the term beguine (*béguin*). Thus, while the religious life, in the words of P. H. Cullum, 'required the setting aside of emblems of masculine authority and autonomy', Robert clearly felt the need to insist on masculine terminology for the secular clergy.[20]

I Masculinity and the secular cleric

By definition, the secular cleric lived in the world, yet he was supposed to remain distinct from the laity he served by virtue of his superior morality and learning. The Gregorian Reform, which imposed celibacy upon the secular clergy, sought to 'monasticize' the clergy as it separated them from laymen and especially all women.[21] Indeed, reform legislation before and beyond the Gregorian Reform sought to construct a clerical masculinity that was in every way different from lay masculinity.[22] Attaining this high standard of behavior and knowledge amongst the clergy was a constant challenge, as evidenced by numerous attempts to reform the secular clergy as well as virulent criticisms of the priesthood.

Scholarship on medieval masculinity and the clergy has primarily focused on the issue of chastity since it was one of the main expressions of masculinity prohibited to clerics.[23] Several studies have questioned the effects of this imposition of celibacy on understandings of clerical masculinity as well as the extent to which violations of celibacy vows might have been an attempt to emulate the masculinity of the laity.[24] While sexuality was certainly an important aspect of masculinity, one that was under constant threat by virtue of the secular cleric's responsibility to provide pastoral care to the laity, it was not the only expression of clerical masculinity.[25] Indeed, several recent studies that address clerical celibacy have argued that clerics did not see themselves as 'emasculine'. Rather, they constructed an alternate code of masculinity that emphasized self-restraint.[26] Secular clerics distinguished their masculinity from that of monks, arguing that contact with the world, which exposed secular clerics to temptation, demonstrated their greater self-mastery. Moreover, they claimed a monopoly over the public offices of preaching and teaching, insisting that monks lacked the authority to preach. These same claims resurfaced with the coming of the mendicant orders in the early thirteenth century. [27]

This conception of clerical masculinity that eschewed sexual activity, violence and other behaviors defined as 'lay' was not hegemonic, among members of the secular clergy. Indeed other factors were perhaps more important in defining clerical masculinity. By the early thirteenth century, the secular clergy was developing a strong collective identity, especially in their pastoral responsibilities. The secular clergy understood its purpose in the ecclesiastical hierarchy as necessitating a religious life in contact with the world rather than enclosed within a monastery. [28] The collective identity of the secular clergy, which was in part formed in opposition to that of enclosed monks, was centered upon the seculars' responsibility to provide pastoral care to the laity.

This sense of identity as distinct from both the laity and the monastic clergy was nowhere more apparent than at the University of Paris.[29] The universities fit into the seculars' sense of their own purpose, not only because they had established guilds of masters separate from monastic schools, but also because the secular clergy's responsibilities to the laity, they believed, were best carried out if informed by university-trained theologians. Parisian masters were particularly conscious of their role in defending and explaining matters of faith and jealously guarded their privileges, since they believed that their ability to determine Christian truth depended on the preservation of these privileges.[30]

University theologians not only constructed their identities in relation to the lower clergy, they also further distinguished their masculinity from that of the laity through their intellectual abilities, defining scholarly work as the 'most masculine' of vocations.[31] By virtue of his presence at the university, the cleric had already proven that he was not a woman. In this single-sex environment, clerics were trained to best their peers in the disputation, where they displayed their knowledge of textual authorities, superior logic and familiarity with the technical vocabulary of scholastic thought. These skills reflected the theologian's years of training, which separated him not just from women, but from irrationality in general, a quality associated with beasts.[32] Though behavior deemed characteristic of lay masculinity certainly had an appeal to some university clerics, the ideal of clerical masculinity – at least among university scholars – was shaped by their unique position and privileges within the university.

II Masculinity and the secular-mendicant conflict

The university itself was founded as a voluntary association of secular teaching masters.[33] This association of secular clerics fought for and protected its privileges by acting as a unified body, calling for a cessation of scholarly activities when its interests were threatened.[34] Over time, members of the university clearly understood their reputations as scholars as dependent upon the preservation of their scholarly privileges, especially protection from physical violence, independence from outside interference and the right to determine its own membership.[35] By the late thirteenth century, the masters of theology possessed an especially elevated notion of the importance of their intellectual expertise. Clearly, secular masters viewed their privileges as worth vigorously fighting for.[36]

In the thirteenth century, the secular clergy's masculine identity on both the parish and university levels was challenged by the coming of the mendicant orders. The Dominicans and Franciscan were founded as preaching orders and saw it as their mission to go out to the urban centers, preaching and hearing confession, tasks over which the secular clergy claimed exclusive responsibility. The mendicants, moreover, represented the ultimate monasticization of the clergy since they lived according to a rule, like monks, but served the laity like secular clerics. As monks who were not enclosed behind monastery walls but wandered from town to town preaching to the laity, the Franciscans and Dominicans seemed to usurp the active duties of the secular clergy while claiming to seek spiritual perfection through the contemplative life. In addition to challenging the secular clergy's pastoral authority, the mendicants attacked the secular clergy's image by criticizing them for their worldliness and reliance on benefices, which contrasted sharply with the mendicants' poverty.[37]

Thus, the differences between monastic and clerical masculinities were called into question by the arrival of the friars. Immensely popular preachers, theologians and confessors, the mendicants threatened the secular clergy's position at all levels. On the parish level, the mendicants posed a threat to local priests because they preached to the laity, heard confessions and competed with secular clergy for the charity of the laity.[38] Moreover, the mendicant orders, because of their vows of poverty, obedience to the pope and superior training programs, appeared to be the better pastors.[39]

Conflict between the seculars and the mendicants was particularly intense at the University of Paris. The seculars initially welcomed the mendicants' arrival in Paris in the early thirteenth century. They soon, however, recognized the threat the friars posed to the preservation of scholarly privileges upon which the secular clergy depended. The mendicants set up their own schools in Paris, which they opened to secular students, and they competed successfully with secular masters for positions within the Church and at the French court, especially as royal confessors and almoners.[40] The mendicants quickly gained reputations at the university as learned and charismatic preachers and teachers. Moreover, the mendicants' theological program worked to undercut the secular clergy, and Franciscan and Dominican friars soon gained a reputation for superior theological learning and attracted secular students to their schools.[41]

Tensions between the orders were exacerbated as mendicant scholars, while participating in university life as teachers and benefiting from the university's reputation for theological learning, refused to protect the university's interests. In 1239, the mendicant orders famously refused to take part in a strike called in response to violent attacks on the university's members. Moreover, the mendicants willfully ignored the secular masters' claims to govern university membership. Over the objections of the secular clergy, the mendicants managed to acquire three chairs on the faculty of theology at the expense of secular clerics.[42]

The conflict between the secular and mendicant orders intensified when the mendicant orders refused to participate in a university-wide strike called in spring of 1253 in response to a brutal assault on four students, one of whom was killed.[43] The mendicants, whose loyalties were to their orders rather than to the university itself, were unwilling to protect the rights of the university in this manner. Thus, in the view of the secular clergy, not only were the mendicants absorbing chairs that had once belonged to the seculars and luring secular students to their schools, they also threatened the existence of the university as an independent body of masters since, by refusing to participate in the strike, the mendicant orders damaged the university's ability to protect its rights and privileges.[44] In response to the mendicant masters' refusal to participate in the strike, the secular members expelled them from the university.[45] The mendicants responded by appealing to the pope, who forced the university to readmit the friars in July 1253, further demonstrating the mendicant threat to the university's ability to function independently.

During the course of this conflict, both groups sought to negotiate conflicting understandings of clerical masculinity in order to defend their own positions as intellectual and pastoral authorities. While the mendicant orders challenged the secular clergy on their worldliness, arrogance and ignorance, the secular clergy called into question the mendicant orders' right to preach and teach (privileges they viewed as exclusive to the secular clergy), as well as the mendicants' vows of poverty. The mendicants benefited from the support of their orders and the protection of the papacy, effectively freeing them from having to concern themselves with the attainment of benefices. Moreover, mendicants were able to claim the moral high ground by virtue of their poverty, portraying themselves as obedient and humble in contrast to the arrogant and greedy secular cleric.

The secular clergy claimed that their way of life was in many ways more heroic than that of monks, who lived in monasteries isolated from the temptations of the world. The secular master, William of Saint-Amour, who emerged as the spokesman for the secular clergy during the height of the secular-mendicant conflict in the 1250s, asserted that mendicants were monks, and thus were not called to preach and hear confession. These responsibilities, he argued, were reserved for the ordained priests, who were the descendants of Jesus' Apostles.[46]

III   Robert of Sorbon and the secular-mendicant conflict

It was within this context of secular-mendicant competition and conflict that Robert of Sorbon founded a college for secular clerics pursuing doctorates in theology. An admirer of the mendicant training program but a firm believer in the necessity of the secular clergy, Robert sought to develop a program at the Sorbonne for secular clerics focused on pastoral care. Thus, in contrast to William of Saint-Amour and Gerard d'Abbeville, who employed the scholastic weapons of quolibets and scholastic treatises in their attacks on mendicant privileges, Robert contended with mendicant competition by focusing his efforts on the pastoral reform of the secular clergy.[47]

In Robert's view, the mendicants' discipline, reputations for humility and austere lifestyles were appropriate to the secular clergy, who were supposed to be committed to the needs of the laity. Scholastic discourse, competition and ambition only harmed the image of the secular clergy, whose spiritual callings were supposed to be oriented toward society and whose discipline depended not on a rule or monastic oversight, but rather upon individual efforts to exhort one another to live exemplary lives.[48] Robert's efforts to reform the secular clergy therefore emphasized the masculinity of the pastoral over the scholastic theologian.

Robert hoped to convince students and masters that theological study and disputation should be for the benefit of the community of believers, not the individual renown of the learned scholar. Robert insisted that students' efforts were best focused on effective pastoral care rather than the understanding of theological subtleties. Thus, Robert's program sought to reorient the priorities of the secular clergy by encouraging them to put aside their pride in their theological knowledge and to focus on applying their learning to the provision of pastoral care.

Robert's numerous sermons and moral treatises, as well as the statutes he composed for the Sorbonne, underlined the importance of pastoral care and virtuous behavior. The Sorbonne's statutes emphasized morality, intellectual ability and diligent preparation for effective pastoral care.[49] Reflecting the preoccupations of their author, the statutes required students to demonstrate a commitment to both preaching and teaching by making 'progress in preaching, preparing sermons, giving disputations and delivering lectures within seven years of being admitted'.[50] In contrast to some of his contemporaries, Robert considered preaching to be the most important task of the secular cleric. The statutes of the Sorbonne required students to take theology courses, attend disputations and preach in the parishes 'in the spirit of charity'.[51] According to the statutes, failure to meet this requirement would result in loss of the college's support.

Significantly, in his efforts to promote this pastoral masculinity, Robert called for the secular clergy to emulate beguines. Here, too, Robert may have been inspired by the mendicant orders' example. The mendicants were popular preachers and teachers, which may have been due in part to their spiritual friendships with inspired women. Recent scholarship on such relationships has emphasized the ways in which male theologians benefitted from their collaborations with religious laywomen, particularly beguines. Thus, while it is evident that university scholars paid little attention to women during the course of academic debate, in their roles as preachers and pastors to the laity, the university-trained clerics forged a clerical masculinity shaped by relationships with or ideas about religious women.[52]

IV   Clerical masculinity and female spirituality

Recent scholarship on male and female spiritualities has illuminated the ways in which clerics sought to differentiate clerical authority from the unofficial spiritual powers ascribed to certain women.[53] John Coakley, for example, has discussed the collaborative relationship between inspired women and learned male theologians and how male clerics hoped to benefit both professionally and personally from their collaborations with such women.[54] The focus on learning and disputation, some argued, not only distracted the scholar from his personal relationship with God, it weakened his ability to preach effectively and provide pastoral care to the laity. In the carrying out of the scholars' pastoral duties, which were gendered masculine, clerics sought inspiration, and even guidance, from women. Women, as several scholars have noted, were also useful in criticizing male pride and ambition.[55] In some cases, the visions and writings of the inspired woman could be used by her confessors to criticize the sinfulness of university colleagues or as a way to defend one order against the attacks of another.[56]

On a personal level, relationships with inspired women could provide the cleric with access to God. Indeed, it was precisely those aspects of clerical life that excluded women, such as university study, preaching and administering the sacraments, that seemed to separate male clerics from God and draw them to religious women. As Coakley puts it 'the very exclusion of women from the realm of priestly authority ironically endowed them with a new significance outside of it.'[57] Coakley has suggested, moreover, that because of their official responsibilities as leaders in the Church, clerics were less likely to experience the same level of intimacy with the divine experienced by their female penitents.[58] The cleric's relationship with an inspired woman, then, was supposed to provide him with a means of overcoming this perceived distance caused, ironically, by his clerical status.

Thus, for some clerics, this association with women was important in the construction of a clerical masculinity that emphasized pastoral care of the laity over scholastic competition. It was, however, particularly risky in light of concerns about clerical celibacy. While friars, who lived in common and cultivated reputations of piety and humility, might have felt that their relationships with religious women were above reproach, these relationships left the friars vulnerable to accusations of immorality. At the height of the secular-mendicant conflict, members of the secular clergy and their supporters targeted this aspect of the mendicants' spiritual authority by casting suspicion on the friars' relationships with women.[59] As is well known, the mendicant orders, especially the Dominicans, tended to provide beguine communities with pastoral care.[60] This was the case in Paris, where the French king entrusted the Dominican Order with pastoral responsibilities over the royal beguinage.[61] Settlement patterns of individual beguine households also suggest that Parisian beguines preferred to associate themselves with the mendicant orders. Households of beguines emerged around the Franciscan and Dominican convents on Paris' Left Bank sometime in the mid-thirteenth century. [62] While we cannot know for certain if the beguines who settled around the Franciscan and Dominican convents actively sought out the mendicants as their confessors, or if they were drawn to these areas by other factors, the relationship between beguines and mendicants in Paris drew the criticism of secular clerics at the University of Paris.

Seeking to undermine the friars' claims to authority on both the parish and university levels, William Saint-Amour called into question the mendicants' morality and humility by portraying the mendicants' relationships with beguines as excessively intimate, even sexual. According to William, the Dominicans preached to the beguines, heard their confessions and cultivated personal relationships with these women by conversing with them and exchanging letters.[63] Such relationships, in William's view, were particularly suspect due to the beguines' youth and beauty as well as the friars' insistence upon meeting with their female penitents in private.[64] In his famous antimendicant tract *De periculis novissimorum temporum*, William associated the friars with the false prophets identified in Paul's Second Letter to Timothy ('For of these sort are they who creep into houses, and lead captive silly women laden with sins, who are led away with diverse desires'). Although William was careful not to identify the friars specifically, the treatise was a thinly veiled attack on the mendicant orders and their associations with women, particularly beguines.[65] William uses this association to accuse the friars of false teachings, and he explicitly mentions sexual intimacy, stating 'in the beginning, perhaps they may seem through hypocrisy to have only a spiritual intimacy with them, but in the end, they are joined with them for the most part sexually.'[66]

Thus, William condemned what he perceived as a mendicant usurpation of the secular clergy's pastoral responsibilities by calling into question the friars' vows of chastity. In William's view, only the secular clergy had the moral fortitude to preach and teach among the laity without falling into sexual temptation.[67] These claims demonstrate that William distinguished the clerical masculinity of the secular clergy from that of both the mendicants and the laity. Although lay understandings of masculinity surely emphasized sex with women, for the clergy out in the world, to indulge in sexual pleasure damaged the cleric's pastoral authority. Indeed, by implication, such behavior could be perceived as feminine.[68] To emulate the laity by engaging in sexual activity was to surrender the moral authority the pastor needed to maintain his position over the laity.

V   Robert of Sorbon and the beguines

Despite, or perhaps because of, the fact that the beguines were typically associated with the spirituality of the mendicant orders, Robert incorporated ideas about beguines in his efforts to reform the secular clergy. Deeply concerned about the worldly preoccupations of university scholars, Robert drew on the example of the beguine in order to advance a model of humility and charity for his students at the Sorbonne, whom Robert believed pursued personal advancement while neglecting their pastoral duties, particularly preaching and hearing confession. By drawing attention to beguines, Robert did not aim simply to shame negligent clerics; rather, he hoped to encourage secular theologians to adapt the beguine life to their own vocation as ministers to the laity.

Like many of his contemporaries, Robert considered university theologians essential to the preservation of orthodoxy. In this view, the disputation was the means by which the theologian determined correct doctrine, which he then imparted to priests and other ministers to the laity. The theologian, however, could not fulfill his responsibilities through study and disputation alone. Rather, he ought to devote his attention to effective preaching and pastoral care, tasks that Robert insisted were best carried out by moral pastors committed to the care of their flock.[69] As several scholars have noted, Robert's sermons and treatises reveal a preoccupation with the health of his students' souls, encouraging contrition, confession, penance and *caritas* among university scholars with the aim of producing more knowledgeable and effective preachers and confessors.[70] What has received less attention is the ways in which Robert used ideas about beguines in his efforts to encourage reform among the secular clergy.

In a sermon Robert preached at the beguinage of Paris in the early 1270s, Robert favorably compared beguines with university scholars.[71] Citing a passage in the Gospel of Matthew (Mt 13:44: 'The kingdom of heaven is like unto a treasure hidden in a field. Which a man having found, hid it, and for joy thereof goeth, and selleth all that he hath, and buyeth that field.'), Robert explained to his audience that the beguinage itself represented the field.[72] While conceding that the beguinage – a structure he portrays as open to the world – was an unlikely place to hide something of such value, Robert asserted that – like the beguinage – the

Kingdom of Heaven is available to anyone willing to humble themselves for the sake of God. Humility, then, is a prerequisite for both the beguinage and heaven.

While a questionable compliment to the beguines in his audience, Robert's sermon is really more concerned with the 'great and wise masters' who shun the beguinage as a place of little worth, overlooking the treasure housed there as a consequence. Describing the beguinage as the most reviled of all communities, Robert's sermon contrasts the beguines to worldly masters who strive for honor and recognition.[73] Although mocked and scorned, the beguines serve God and one another. The 'great and wise masters of the world', however, value their reputations above all else, declaring '*Fi de béguinage*! I want to be a *prud'homme* but not a *béguin*.'[74] The reviled beguine, then, possesses God's treasure by virtue of her willingness to endure scorn. The esteemed *prud'homme*, because of his preoccupation with reputation and personal advancement, overlooks the treasure, which is right before his eyes.[75]

Robert's sermon at the beguinage, which was likely delivered before an audience of beguines and university clerics, aimed to critique his secular colleagues' preoccupation with career advancement by comparing them unfavorably with beguines. Indeed, contrasting the *prud'homme* with the *béguin* seems to have been a significant theme in Robert's writings. It is through this comparison with the *prud'homme*, who represents the esteemed theologians, that Robert advanced a definition of the term *béguin* that emphasized and embraced the negative reactions some of his contemporaries had to lay religiosity while moving away from an exclusive association of the term with female lay religiosity.

According to Robert, *béguine* (and its male form *béguin*) were terms derisively applied to women and men who confessed frequently, heard the word of God eagerly and corrected their fellow Christians unfailingly.[76] In short, they behaved in a manner that observers found too pious to be genuine, and indeed Robert declares that no community was more reviled than the beguinage.[77] The writings of some of Robert's contemporaries, especially William of Saint-Amour and his supporters, seem to support this claim. Robert, however, embraced the negative connotations of the term in order to encourage university clerics to practice humility and show love for one another (in other words *caritas*, an important theme in Robert's writings).[78] According to Robert, beguines endured skepticism and accusations of hypocrisy with patience, refusing to tone down their practices, even in the face of ridicule. Robert thought this sort of patience, humility and integrity was particularly important for the secular clergy since they, like the beguines, were a visible presence in the city and their actions attracted much scrutiny. In his popular treatise *On Conscience*, Robert notes that some men (and here he is likely referring to members of the secular clergy), conscious of this scrutiny, behaved one way in front of religious men and another in the presence of laymen, 'deriding béguines and religious men to gain the good-will of laymen'.[79] Beguines, according to Robert, cared little for the opinions of others.

By pairing the terms *prud'homme* and *béguin*, Robert urged his listeners to focus on their interior selves rather than their public reputations, an issue that, as Clare Waters has shown, was of particular concern for preachers.[80] This issue of public reputation is evident in Robert's defense of the *béguin* in the course of a debate with Jean of Joinville in the presence of King Louis IX.[81] The disagreement, which Joinville describes in his *Life of Saint Louis*, reveals that the contrast between the *prud'homme* and the *béguin* was a major theme in Robert's thought. *Prud'homme*, a word used in courtly literature to refer to the ideal knight and later applied to men recognized as experts in their professions, designated someone of esteem and good reputation.[82] The term *béguin*, as is well known, had much more negative connotations, the most evident being its associations with a type of spirituality characterized as feminine.[83] When applied to university clerics, these terms suggest two different understandings of clerical masculinity.

In Robert's writings, as well as the debate with Joinville, the terms seem to refer to subjective valuations rather than distinct religious identities. As Nicole Bériou has noted, the terms signify two different ways of 'being named, and therefore judged, by others'.[84] They also suggest two possible codes of masculine behavior. By advocating the *béguin* over the *prud'homme*, Robert asserted his preference for a clerical identity that emphasized humility and accessibility. In his praise for the *béguin* in his sermons and moral treatises, Robert clearly hoped to encourage the secular clergy to reject the aggressive, competitive masculinity measured by the admiration of others. Indeed, in his sermon comparing the *prud'homme* and the *béguin*, Robert argued that it was the *béguin's* disregard for reputation and willingness to endure scorn that earned him salvation. The *prud'homme*, however, was unwilling to pay this price, which Robert asserted was necessary to obtain the Kingdom of Heaven.

By contrasting the beguines' simple humility with the clerics' preoccupation with worldly success, Robert's sermon raises issues that, as we have seen, were of particular concern for contemporary understandings of clerical masculinity in the context of the secular-mendicant controversy. In Robert's view, preoccupation with scholastic learning and competition with other men undermined the secular clergy's ability to provide effective pastoral care to the laity and, more importantly, imperiled their immortal souls.[85]

In his sermons and moral treatises, Robert encouraged clerical morality by drawing attention to the *béguin*, who, according to Robert, scrupulously examined his conscience, confessed frequently and eagerly attended sermons. In his treatise *On Conscience*, Robert compared the Last Judgment to university examinations, reminding students that they would not dare go before the Chancellor of the University without reading the book on which they would be examined. Likewise, no one should go before God without having first examined thoroughly the book of his conscience. In this treatise, Robert praised the *béguins* and *béguines* for their understanding of the 'Book of Conscience', which they acquired through frequent confession.[86]

Robert was not alone in associating beguines with a sacramental piety focused on confession and penance. The thirteenth-century hagiographies collectively referred to as the 'béguine *vitae*' describe women who thoroughly confess their sins, however minor. In the prologue to his influential *Life of Mary of Oignies* the Paris-educated cleric James of Vitry (1170–1240) described the conscientiousness of women he observed in the diocese of Liège. Addressing the bishop of Toulouse, who came to Liège after being exiled from his see by the Cathars, James recounts how the bishop was 'filled with wonder by certain women who wept more for a single venial sin than the men of your lands over a thousand mortal sins'.[87] The beguines' scrupulosity was also described in negative terms. For example, the Dominican Minister General Humbert of Romans complained about beguines who confessed too frequently.[88] Modern studies of female spirituality have noted the beguines' confessional and penitential scrupulosity, describing frequent confession as a characteristic of 'female spirituality'.[89] Unlike many of his contemporaries, Robert did not praise the confessional practices of female penitents with the aim of demonstrating the confessors' sacramental abilities. Nor did he criticize the confessor–penitent relationship in order to cast suspicion on the mendicants' associations with beguines. Rather, Robert hoped to instruct his students to *emulate* the beguines' confessional practices. In other words, Robert sought to claim this 'feminine' spiritual practice for the secular clergy.

It is difficult to know the extent to which Robert's understanding of the term *béguin* was affected by contact with actual women living at the beguinage. In the sermon locating the Kingdom of Heaven in the beguinage, Robert emphasized the institution's openness to the world, a possible reference to the rules of the house, which allowed residents to enter and leave and receive visitors.[90] Robert's reference to this openness brought him to one of the central themes in the sermon: the beguines' particularly active role in exhorting their fellow Christians in the world. The beguines' defining virtue, according to Robert, was their *caritas*, which they exhibited in their fervent efforts to convert sinners.[91] Elsewhere, Robert encouraged clerics to emulate the beguines' *caritas*, which he considered the most important of all virtues. Indeed, Robert frequently referred to the beguines' *caritas* in his sermons and moral treatises. In one such sermon to a university audience, Robert referred to the community as an '*ordo caritatis*' over which Jesus was abbot.[92] In another, Robert put forth theories on the meaning of the term '*béguine*', claiming that the term itself evolved from the words '*bene igne*', which he interpreted as referring to the way in which the *béguines* burned with the fire of *caritas*.[93] According to Robert, university scholars would do well to emulate the beguines in their relations with one another. In a sermon that was likely addressed to students of the Sorbonne, Robert expounds on the theme of *caritas* by referencing beguines. In this sermon, Robert lamented those who ignored their neighbor's sins, commenting that this behavior was not '*debonaire*', but cruel. When they neglected to reprimand and pray for their colleagues, Robert asserted, these students were like the man who did not warn his neighbor that his house was on fire.[94]

For Robert, one of the primary purposes of encouraging contrition, confession, penance and *caritas* among university scholars was to produce more knowledgeable and effective preachers and confessors.[95] Robert's program sought to reorient the priorities of the secular clergy by encouraging them to put aside their pride in their theological knowledge and to focus on applying their learning in the provision of pastoral care. In his efforts to promote this reorientation of priorities, Robert frequently employed the example of the *béguine*. In Robert's view, preoccupation with scholastic learning and competition with other men undermined the secular clergy's ability to provide effective pastoral care to the laity and, more importantly, imperiled their immortal souls. Directly contrasting the beguine with the learned theologian, Robert quotes the Gospel of Matthew, saying 'you have hidden these things from the wise and revealed them to the little ones.'[96] Robert follows this comparison with an admonition to 'those great and wise masters' who think they can know God through study and reason, comparing them to *béguins* and *béguines* who understand that they can know God only through Grace.[97]

Robert's use of both the term *béguin* and *béguine* in his sermon is significant. In Robert's view, secular clerics should emulate the humility and charity of the beguines, who lived and served in the world rather than within the enclosure of a monastery. In this way, Robert was not suggesting that secular clerics associate with beguines because these women possessed spiritual powers that the clerics lacked; rather, Robert urged university scholars to become *béguins* themselves. Robert's frequent references to *béguins*, the masculine form of the term *béguine*, indicate that he saw the beguine life as one that could be profitably adapted for the secular clergy, but he did not wish to apply a feminine term to these men. Robert's representations of the beguine life emphasized a set of behaviors that he wished to encourage among the male secular clergy. In this sense, he denied that the term referred solely to religious laywomen, and – perhaps – in this way disassociated the term from women close to the mendicant orders.

VI Imitation or collaboration?

Robert's efforts to present beguines as models for university clerics presented some challenges to clerical masculinity. Robert portrayed the beguines as women who lived religious lives despite the temptations and criticisms of the world. The university, however, was an environment in which clerical manliness was molded through competition, specifically in the realm of disputation, with other men. Moreover, secular theologians had developed a keen sense of the need to fight for their privileges, which they viewed as inextricably linked to their ability to preserve orthodoxy through their expertise as theologians. Humility and *caritas*, then, were in some ways incompatible with scholastic understandings of clerical masculinity.

Even Robert's students did not adopt his broad, gender-inclusive definition of *béguin*, instead using the term strictly to describe lay religious women. Rather than imitate the *béguin*, Robert's students sought to associate themselves with *béguines*, whom they understood as women fundamentally different from themselves. Thus, in the context of controversies over preaching authority and pastoral care, secular clerics chose collaboration over imitation. Highlighting their own institutionally based authority, secular clerics approached their beguine audiences with certain assumptions about the women's spiritual abilities.

Clerics studying and teaching in thirteenth-century Paris had ample opportunities to interact with women who self-identified as beguines. Beguine households were located near the Sorbonne and the beguinage was a popular place for Parisian students and masters in theology to hone their preaching skills.[98] Clerics affiliated with the Sorbonne frequently attended sermons preached to the beguines, which they preserved in reportations, that is, records of the sermons as they were preached.[99]

These reportations were sometimes compiled into sermon collections and serve as important sources for the influence Robert's teachings had on his students. One of the most complete compilations of recorded sermons was those put together by the Sorbonne cleric, Pierre de Limoges, and significantly – as Nicole Beriou has shown – this collection contains far more sermons preached at the beguinage than at any other place of worship in and around Paris.[100] Indeed, these sermons were copied, compiled and consulted by clerics affiliated with the Sorbonne, indicating a keen interest among secular clerics in sermons preached before the city's beguine communities.[101]

The final page of one collection bequeathed to the Sorbonne by Pierre de Limoges contains a list, in French, of characteristics attributed to beguines. This list, entitled 'The 32 properties of the beguinage', portrays the beguines expressing their love for Christ with 'praying mouths, weeping eyes and desiring hearts'. With 'cherubic understanding and seraphic insight', 'burning love' and 'spiritual courtliness', the beguines 'die by living, live by dying, fast by feasting, feast by fasting'.[102] In addition to this description, which draws on themes familiar to modern scholars of female spirituality, the manuscript contains several *exempla* concerning beguines, including one warning 'one cannot be a good preacher or confessor unless one supports the béguinage.'[103]

One particularly detailed *exemplum* within this collection describes a conversation between a Parisian master and a beguine, whom the master finds weeping in a church out of love for her spouse, whom she identifies as Christ.[104] The master marvels that the beguine, who understands nothing of theological mysteries, could experience God so intimately, when he himself feels nothing. In a reversal of the male cleric/female penitent relationship, the master asks the beguine to explain to him how he might be able to achieve similar desire for God. In response, the beguine explains that it is the cleric's constant study of Scripture that prevents him from knowing God, explaining 'because you see many mirrors in Sacred Scripture, upon which you gaze, your affection (*affectum*) is impeded.'[105]

Thus, despite Robert's attempts to promote the *béguin* as a model for the secular clergy, his students preferred to represent the *béguine* as fundamentally different from themselves. The *exemplum* contrasts the beguine's desire for Christ (*affectus*) and the master's dispassionate reason. The beguine's experiential knowledge is compared with the cleric's text-based knowledge, affirming the master's intellectual authority while expressing some concern about the limits of this knowledge, which the master hoped to supplement with the beguines' experiences. The conversation between the beguine and the Parisian master, however, suggests that the master does not fill a void in his spiritual life by acting merely as a witness to her religious experiences. Rather, he learns to fill this void himself by moving away from a clerical masculinity concerned with scholastic disputation and devoting more attention to his own spiritual progress. The beguine serves as a useful model to the master, but he does not become a *béguin*.

Clearly, clerical discussions of beguines reveal the importance of ideas about lay religious women in the construction of clerical masculinity and university identity. Just as medieval intellectuals sometimes employed imaginary female personifications of virtues in their theological writings, Robert of Sorbon used ideas about the female *béguine* in his sermons and treatises aimed at other men.[106] For Robert, the *béguine* was more than just a model for the secular cleric. She was a useful focus through which Robert negotiated the competing codes of behavior associated with his – and his students' – clerical identity. Yet, as Robert encouraged the secular clergy to emulate or become *béguins*, the *béguine* herself is almost pushed aside, possibly blunting any positive effects Robert's praise might have had on 'real' *béguines*.[107] Robert's contemporaries also seemed inclined to 'think with *beguines*'; however, they chose to see the *béguine* as a religious laywoman with qualities different from, but perhaps complementary to, those of the university cleric. Thus, clerics affiliated with Robert's school seem to have preferred to incorporate rather than emulate beguines when constructing and promoting a masculine spirituality focused on preaching and pastoral responsibility. They sought to preserve the image of the cleric as a rational intellectual authority fundamentally different from women. University clerics established close ties with Paris' beguine community, but university competition and conflict prevented them from adapting the *béguine* life for the secular clergy.

Acknowledgments

I would like to thank Nancy McLoughlin, Andrew Miller and Katrin Sjursen for reading and commenting on earlier drafts of this essay. I owe special thanks to Jennifer Thibodeaux for her helpful suggestions.

Notes

1. Ruth Mazo Karras, *From Boys to Men: Formations of Masculinity in Late Medieval Europe* (Philadelphia, 2003), 75.
2. Ibid.
3. Ian P. Wei, 'The Self-Image of the Masters of Theology at the University of Paris in the Late Thirteenth and Early Fourteenth Centuries', *Journal of Ecclesiastical History* 46 (1995): 398–431.
4. Sharon Farmer, *Surviving Poverty in Medieval Paris: Gender, Ideology, and the Daily Lives of the Poor* (Ithaca, NY, 2002), 44–8.
5. On theological debates over preaching rights, see Alcuin Blamires, 'Women and Preaching in Medieval Orthodoxy, Heresy, and Saints' Lives', *Viator* 26 (1995): 135–52 and Nicole Bériou, 'The Right of Women to Give Religious Instruction in the Thirteenth Century', in *Women Preachers and Prophets through Two Millennia of Christianity*, ed. Beverly Mayne Kienzle and Pamela J. Walker (Berkeley, CA, 1998), 135–45.
6. For an overview of this 'secular-mendicant' conflict, see Yves M.-J. Congar, 'Aspects ecclésiologiques de la querelle entre mendicants et séculiers dans la seconde moitié du XIIIe et le début du XIVe', *Archives d'histoire doctrinale et littéraire du moyen âge* 28 (1961–1962): 35–151.
7. Astril L. Gabriel, 'The Ideal Master of the Medieval University', *Catholic Historical Review* 60 (1974): 1–40.
8. On the quodlibetal disputation as an expression of clerical masculinity, see Karras, *Boys to Men*, chapter 3. On the view that competition among masters endangered the cleric's soul, see Gabriel, 'Ideal Master', 28–30 and Edward T. Brett, *Humbert of Romans: His Life and Views of Thirteenth-Century Society* (Toronto, 1984), 151–66.
9. Gabriel, 'Ideal Master', 29–30 and Wei 'Self-Image', 419–20.
10. Clare M. Waters, *Angels and Earthly Creatures: Preaching, Performance, and Gender in the Later Middle Ages* (Philadelphia, 2004), 48.
11. Katherine Jansen argues that, from their very beginnings, the mendicant orders espoused humility and obedience in addition to poverty, positioning themselves as 'a feminine anti-type to the masculinized Church'. See Katherine L. Jansen, *The Making of the Magdalen: Preaching and Popular Devotion in the Later Middle Ages* (New Jersey, 2000), 85.
12. P.H. Cullum, 'Clergy, Masculinity, and Transgression in Late Medieval England' in *Masculinity in Medieval Europe*, ed. D.M. Hadley (New York, 1999), 178–196; Jennifer Thibodeaux, 'Man of the Church or Man of the Village? Gender and the Parish Clergy in Medieval Normandy', *Gender and History* 18/2 (2006): 380–99; and Ruth Mazo Karras, 'Thomas Aquinas' Chastity Belt: Clerical Masculinity in Medieval Europe', in *Gender and Christianity in*

Medieval Europe, ed. Lisa M. Bitel and Felice Lifshitz (Philadelphia, 2008), 52–67.

13. For an excellent discussion of these relationships, see John W. Coakley, *Women, Men, and Spiritual Power: Female Saints and Their Male Collaborators* (New York, 2006).

14. Karras, 'Using Women to Think Within the Medieval University', in *Seeing and Knowing: Women and Learning in Medieval Europe 1200–1550*, ed. Anneke Mulder-Bakker (Brepols, 2004), 21–33.

15. On Robert of Sorbon, see Palémon Glorieux, *Aux origines de la Sorbonne*, volume 1, *Robert de Sorbon, l'homme, le collège, les documents* (Paris, 1965–1966); Astrik L. Gabriel, *The Paris Studium: Robert of Sorbonne and his Legacy* (Notre Dame, IN, 1992); and Nicole Bériou, 'Robert de Sorbon', in *Dictionnaire de spiritualité: ascétique et mystique, doctrine et histoire*, ed. André Vauchez, 13 (1988): col. 816–824.

16. For a good summary, see Walter Simons, *Cities of Ladies: Beguine Communities in the Medieval Low Countries, 1200–1565* (Philadelphia, 2001), especially chapter 5. Ernest McDonnell's work on beguines communities also remains influential; see *The Beguines and Beghards in Medieval Culture, With Special Emphasis on the Belgian Scene* (New Brunswick, N.J., 1954). For official, canonical interpretations of the beguine status see Elizabeth Makowski, *'A Pernicious Sort of Woman': Quasi-Religious Women and Canon Lawyers in the Later Middle Ages* (Washington, D.C., 2005). For an insightful survey of beguine historiography, see Jennifer K. Deane, 'Beguines Reconsidered: Historiographical Problems and New Directions', *Monastic Matrix* (2008), Commentaria 3461, available at http://monasticmatrix.org/commentaria/article.php?textId=3461.

17. For specific references to such attacks, see Herbert Grundmann, *Religious Movements in the Middle Ages*, trans. Steven Rowan (Notre Dame and London, 1995), 139–52; McDonnell, *The Beguines and Beghards*, 430–73, and Tanya Stabler Miller, 'What's in a Name? Clerical Representations of Parisian Beguines (1200–1328)', *Journal of Medieval History* 33/1 (2007): 60–86.

18. Grundman, *Religious Movements*, 141–2.

19. Caroline Walker Bynum, 'Jesus as Mother and Abbot as Mother: Some Themes in Twelfth-Century Cistercian Writing', in *Jesus as Mother: Studies in the Spirituality of the High Middle Ages* (Berkeley, CA, 1982), 110–69, and Katherine L. Jansen, *The Making of the Magdalen: Preaching and Popular Devotion in the Later Middle Ages* (Princeton, NJ, 2000), 84–6.

20. P.H. Cullum, 'Introduction: Holiness and Masculinity in Medieval Europe', *Holiness and Masculinity in the Middle Ages*, ed. P.H. Cullum and Katherine J. Lewis (Cardiff, 2004), 4.

21. Jo Ann McNamara, 'The *Herrenfrage:* The Restructuring of the Gender System, 1050–1150', in *Medieval Masculinities: Regarding Men in the Middle Ages*, ed. Clare A. Lees (Minneapolis, MN, 1994), 3–29.

22. Maureen Miller has examined discourses in the reforming church that constructed a clerical masculinity distinct from lay masculinity. See Maureen C. Miller, 'Masculinity, Reform, and Clerical Culture: Narratives of Episcopal Holiness in the Gregorian Era', *Church History* 72 (2003): 25–52.

23. See especially P.H. Cullum, 'Clergy, Masculinity, and Transgression'; Jacqueline Murray, 'Masculinizing Religious Life: Sexual Prowess, the Battle for Chastity, and Monastic Identity', in *Holiness and Masculinity in the Middle Ages*, 24–42; and Karras 'Thomas Aquinas's Chastity Belt: Clerical Masculinity in Medieval Europe', in *Gender & Christianity in Medieval Europe: New Perspectives*, ed. Lisa Bitel and Felice Lifshitz (Philadelphia, University of Pennsylvania Press, 2008), 52–67.

24. On the effect of such vows on the gender identity of clerics, see the opposing views of Thibodeaux, 'Man of the Church', and R. N. Swanson, 'Angels Incarnate: Clergy and Masculinity from Gregorian Reform to Reformation', in *Masculinity in Medieval Europe*, 160–77. See also studies cited above.

25. As Jennifer Thibodeaux has argued, medieval concepts of masculinitity were 'not universally based on one set of criteria'. See Thibodeaux, 'Man of the Church', 381.

26. On clerical 'emasculinity', see Swanson, 'Angels Incarnate'. On the notion of masculinity as self-mastery see Jacqueline Murray, 'Masculinizing Religious Life' and Karras 'Thomas Aquinas's Chastity Belt', cited above.

27. M. Peuchmaurd, 'Mission canonique et prédication: Le prêtre ministre de la parole dans la querelle entre mendiants et séculiers au XIIIe siècle', *Recherches de théologie ancienne et médiévale* 30 (1963): 122–44, 251–76.

28. Francis Rapp, 'Rapport Introductif', *Le clerc séculier au moyen âge* (Paris, 1993) 1–15.

29. On the secular clergy's collective identity, see Ian P. Wei 'The Masters of Theology at the University of Paris in the Late Thirteenth and Early Fourteenth Centuries: An Authority Beyond the Schools', *Bulletin of the John Rylands University Library of Manchester*, 75/1 (1993): 37–63. Here, Wei points out that the prelates reminded Parisian masters of their shared interests, specifically that the concerns of the prelates and diocesan clergy were the same as those of the masters. On the relationship between masters and the lower clergy, see Alan E. Bernstein, 'Magisterium and License: Corporate Autonomy against Papal Authority in the Medieval University of Paris', *Viator* 9 (1978): 291–308. For an insightful discussion of the secular clergy's collective identity in the context of secular-mendicant conflict, see Nancy McLoughlin, '*Universitas*, Secular-Mendicant Conflict and the Construction of Learned Male Authority in the Thought of John Gerson (1363–1429)' (unpublished PhD dissertation, University of California, Santa Barbara, 2005).

30. Bernstein, 'Magisterium and License', 301.

31. Farmer, *Surviving Poverty*, 44–8.

32. Karras, *Boys to Men*, 68.

33. On the origins of the University of Paris, see Stephen C. Ferruolo, *The Origins of the University: The Schools of Paris and their Critics* (Stanford, CA, 1985).

34. Pearl Kibre, *Scholarly Privileges in the Middle Ages: The Rights, Privilege, and Immunities of Scholars at Bologna, Padua, Paris, and Oxford* (Cambridge, MA, 1962), 85–131.

35. See Alan Bernstein, 'Magisterium and License' and *Pierre d'Ailly and the Blanchard Affair: University and Chancellor of Paris at the Beginning of the Great Schism* (Leiden, 1978). For an excellent discussion of these associations in

the context of secular-mendicant conflict during the fourteenth century (though drawing on the origins of these associations in the thirteenth century), see Nancy McLoughlin, 'Gerson as Preacher in the Conflict between Mendicants and Secular Priests', in *A Companion to Jean Gerson*, ed. Brian Patrick McGuire (Leiden, 2006), 249–91.

36. Stephen C. Ferruolo, '*Parisius-Paradisus*: The City, Its Schools, and the Origins of the University of Paris', in *The University and the City: From Medieval Origins to the Present*, ed. Thomas Bender (Oxford, 1988), 22–43.

37. On this conflict, see Decima L. Douie, *The Conflict Between the Seculars and the Mendicants at the University of Paris in the Thirteenth Century* (London: 1954); Michel-Marie Dufeil, *Guillaume de Saint-Amour et la polémique universitaire parisienne,1250–1259* (Paris, 1972).

38. On the mendicant orders and secular response, see David L. d'Avray, *The Preaching of the Friars: Sermons Diffused From Paris Before 1300* (Oxford, 1985), 49–53 and D.H. Lawrence, *The Friars: The Impact of the Early Mendicant Movement on Western Society* (London and New York, 1994).

39. David L. D'Avray, *The Preaching of the Friars*, 49.

40. Xavier de La Selle, *Le service des âmes à la cour: confesseurs et aumôniers des rois de France du XIIIe au XVe siècle* (Paris, 1995).

41. D'Avray, *Preaching of the Friars*, 50.

42. Hastings Rashdall, *The Universities of Europe in the Middle Ages*, ed. F. M. Powicke and A. B. Emden (Oxford, 1936), volume one, 372–7.

43. Dufeil, *Guillaume de Saint-Amour*, 5.

44. Bernstein, 'Magisterium and License', 300–3.

45. See McLoughlin, '*Universitas*', 113.

46. Douie, *The Conflict Between the Seculars and the Mendicants* and Penn Szittya, *Antifraternal Tradition*, 44–6. See also, Yves M.-J. Congar, 'Aspects ecclésiologiques', 35–151.

47. On William of Saint Amour and Gerard d'Abbeville's approach to resolving the conflict, see Douie, *The Conflict Between the Seculars and the Mendicants*, 12–30.

48. Francis Rapp, 'Rapport introductif', in *Le clerc séculier au Moyen Age* (Paris, 1993), 9–25.

49. Gabriel, *The Paris Studium*, 80.

50. Ibid, 85.

51. Bériou, *L'avènement des maîtres de la Parole*, 85.

52. For the argument that university clerics rarely mention women in the course of academic debates, see Ruth Mazo Karras, 'Using Women to Think With'.

53. See, for example, *Gendered Voices: Medieval Saints and their Interpreters*, ed. Catherine Mooney (Philadelphia, PA, 1999); Dyan Elliott, *Proving Woman: Female Spirituality and Inquisitional Culture in the Later Middle Ages* (Princeton, NJ, 2004); and most recently John Coakley, *Women, Men, and Spiritual Power*.

54. See especially John Coakley, 'Gender and Authority of the Friars: The Significance of Holy Women for Thirteenth-Century Franciscans and Dominicans', *Church History* 60/4 (1991): 445–60; 'Friars as confidants of holy women in medieval Dominican hagiography', in *Images of Sainthood in*

Medieval Europe, ed. Renate Blumenfeld-Kosinski and Timea Szell (Ithaca, 1991), 222–46; and *Women, Men, and Spiritual Power*.

55. Bynum, 'Women's Stories, Women's Symbols: A Critique of Victor Turner's Theory of Liminality', in *Fragmentation and Redemption: Essays on Gender and the Human Body in Medieval Religion* (NY, 1991), 27–51. See also Coakley, *Women, Men, and Spiritual Power*.

56. This seems to have been particularly true of the mendicant orders, whose members often acted as the hagiographers and spiritual advisors of religious laywomen. See Coakley, 'Friars as Confidants of Holy Women in Medieval Dominican Hagiography', in *Images of Sainthood in Medieval Europe*, ed. Renate Blumenfeld-Kosinski and Timea Szell (Ithaca and London, Cornell University Press, 1991), 222–46 and 'Gender and Authority of the Friars: The Significance of Holy Women for Thirteenth-Century Franciscans and Dominicans', *Church History* 60 (1991): 445–60.This was certainly the case with Mechthild of Magdeburg and the Dominican order. See Sara S. Poor, *Mechtild of Magdeburg and her Book: Gender and the Making of Textual Authority* (Philadelphia, 2004).

57. Coakley, *Women, Men, and Spiritual Power*, 2.

58. Coakley, 'Friars, Sanctity, and Gender: Mendicant Encounters with Saints, 1250–1325', in *Medieval Masculinities: Regarding Men in the Middle Ages*, ed. Clare A. Lees (Minneapolis, MN, 1994), 91–110.

59. These attacks have received much attention in the literature on the secular-mendicant conflict, though without tying the attacks specifically to clerical masculinity. See Herbert Grundmann, *Religious Movements*, 141–42 and Penn R. Szittya, *The Antifraternal Tradition in Medieval Literature* (Princeton, NJ, 1986), 59–60. My thinking is greatly influenced here by Nancy McLoughlin's analysis of similar strategies employed by secular clerics against the mendicants in the fifteenth century. See McLoughlin, '*Universitas*', especially chapter 5.

60. John Freed, 'Urban Development and the 'Cura Monialium' in Thirteenth-Century Germany', *Viator* 3 (1972) 311–27; Grundmann, *Religious Movements in the Middle Ages*, 139–52; McDonnell, *Beguines and Beghards*, 196–204.

61. Léon Le Grand, 'Les béguines de Paris', *Mémoires de la société de l'histoire de Paris et de l'Ile-de-France* 20 (1893): 350.

62. Sharon Farmer, 'Down and Out and Female in Thirteenth-Century Paris', *American Historical Review* 103/2 (1998): 345–72 and Tanya Stabler 'Now She is Martha, Now She is Mary' (unpublished PhD dissertation, University of California, Santa Barbara, 2007), especially chapter 2.

63. *Collectiones catholicae et canonicae Scripturae*, in *Opera omnia* (Constance, 1632), 267, cited in Grundmann, *Religious Movements*, 141.

64. Penn R. Szittya, *The Antifraternal Tradition*, 60.

65. McDonnell, *Beguines and Beghards*, 456–64; Michel-Marie Dufeil, *Guillaume de Saint-Amour*, 22.

66. *Collectiones*, 196, translation in Szittya, *Antifraternal tradition*, 59.

67. Ibid.

68. Jacqueline Murray, 'Masculinizing Religious Life' and Karras, 'Thomas Aquinas' Chastity Belt', 65.

69. Gabriel, *Paris Studium*, 85.

70. Gabriel, *The Paris Studium*, 80–5 and Bériou, 'Robert de Sorbon', col. 822–24.
71. Nicole Bériou has edited this sermon in an article that analyzes medieval understandings of the terms *béguin* and *prud'homme*. She does not, however, discuss these labels in terms of gender. See Nicole Bériou, 'Robert de Sorbon: Le prud'homme et le béguin', *Académie des inscriptions et belles-lettres* (1994): 469–511.
72. 'Simile est regnum celorum thesauro abscondito in agro, Mt. XIII, quem cum inuenit homo uadit, uendit omnia que habet et emit agrum illum', (BNF Lat. 16507), cited in Bériou, 'Le prud'homme et le béguin', 486.
73. '[I]nter omnes ordines seu religiones plus despicitur beguinagium vel vilipenditur', cited in Bériou, 'Le prud'homme et le béguin'.
74. 'Etiam magni magistri, immo dicunt: 'Fi de beguinagio.' Dicunt enim: 'Bene volo esse probus homo, sed nequaquam beguinus', cited in Bériou, 'Le prud'homme et le béguin'.
75. 'isti magni magistri et sapientes mundani hoc non possunt cognoscere nec istum thesaurum in beguinagio invenire...', cited in Bériou, 'Le prud'homme et le béguin'.
76. 'Sed inter omnes hominess de mundo, qui libentius uadunt ad confessionem et qui libentius audiunt uerbum Dei sunt beguini et beguine', Bériou, 'Le prud'homme et le béguin'.
77. See note 73 above.
78. Nicole Bériou identifies *caritas* as a particularly important theme in Robert's writings; see Bériou, 'Robert de Sorbon', col. 822–3.
79. 'Et quando sunt cum mundanis, faciunt ut mundane, vel pejus, et truphant de beguines et religiosis viris, et derident, ut habeant benevolenciam mundanorum', Robert de Sorbon, *De Consciencia et de tribus dietis*, ed. Félix Chambon (Paris: 1902), 15–16.
80. Waters, *Angels and Earthly Creatures*, 2–9. Of course, it is interesting that Robert's denunciation of hypocrisy praises the *béguin*, a term associated with hypocrisy. See Robert E. Lerner, *The Heresy of the Free Spirit in the Later Middle Ages* (Notre Dame, Ind.: 1993), 39–42.
81. Jean of Joinville, *Life of Saint Louis*, trans. Margaret R.B. Shaw, in *Chronicles of the Crusades* (New York, 1982), 170. Shaw translated *béguin* as 'friar'.
82. Beriou, 'Prud'homme', 474–82.
83. Simons, 121–32.
84. Bériou, 'Prud'homme...', 469.
85. Robert frequently noted the university scholars' lack of concern for their fellow students as well as their neglect of their own personal relationship with God; see Bériou, 'Robert de Sorbon', col. 822–4.
86. *De Conscientia*, 24. Similarly, in his sermon at the beguinage, Robert asserted that the 'béguins and béguines' distinguish themselves by their frequent confession and attendance at sermons.
87. James of Vitry, *The Life of Marie d'Oignies*, trans. Margot H. King (Saskatoon: 1986), 36.
88. Edward T. Brett, *Humbert of Romans. His Life and Views of Thirteenth-Century Society* (Toronto, 1984), 67.
89. For the characterization of women's spirituality in general, see Caroline W. Bynum, *Holy Feast and Holy Fast: The Religious Significance of Food to Medieval*

Women (Berkeley, CA: 1987); and Bynum, *Fragmentation and Redemption*. For its application to beguines in particular, see the essays by Brenda M. Bolton, 'Mulieres Sanctae', in *Women in Medieval Society*, ed. Susan Mosher Stuard (Philadelphia: 1976), 141–56 and 'Vitae matrum: A Further Aspect of the Frauenfrage', in *Medieval Women: Dedicated and Presented to Professor Rosalind M. T. Hill on the Occasion of her Seventieth Birthday*, ed. Derek Baker (Oxford, 1978), 253–73. See also Simone Roisin, 'L'efflorescence cistercienne et le courant féminin de piété au XIII siècle', in *Revue d'histoire ecclésiastique* 39 (1943), 342–78 and the essays in Jocelyn Wogan-Browne, ed., *New Trends in Feminine Spirituality: The Holy Women of Liège and their Impact* (Turnhout, 1999).

90. Royal accounts and property records demonstrate close relations and frequent interaction between residents of the beguinage and outsiders. See Stabler, 'Now She is Martha, Now She is Mary.' The statutes drawn up in 1327 allowed beguines to leave the beguinage to conduct business and visit friends. The Paris beguinage's statutes are housed in the Archives Nationales (AN JJ 73, fol. 52v, no. 71) and were edited by Le Grand, 'Les béguines de Paris', 354–7. Other Northern European beguinages permitted similar traffic in and out of the beguinage. See Simons, *Cities of Ladies*, chapter 3.

91. '[I]nter alios ipsi et ipse habent feruens desiderium ad Deum et feruentiores sunt in conuersione peccatorum quam alii' BNF Lat. 16507, ed. Nicole Bériou, 'Robert de Sorbon: Le prud'homme et le béguin', 490.

92. BNF 15971, 72v, cited in Bériou, 'Robert de Sorbon, Le prud'homme et le béguin', 481, fn. 41.

93. Ibid.

94. Ibid.

95. Bériou, 'Robert de Sorbon', 822–3.

96. BNF Lat. 16507 (trans. Douay-Rheims).

97. 'Et que est causa quare isti magni magistri et sapientes mundani hoc non possunt cognoscere nec istum thesaurum in beguinagio invenire? Hes est ratio, quia non per studium nec per rationem potest iste thesaurus inueniri, sed per gratiam', BNF Lat. 16507, ed. Bériou, 'Le prud'homme', 491.

98. On the locations of beguine households, see Farmer, *Surviving Poverty*, 144–6 and Stabler, 'Now she is Martha', Appendix B. On the presence of clerics at sermons preached at the beguinage, see Nicole Bériou, 'La prédication au béguinage de Paris pendant l'année liturgique 1272–1273', *Recherches augustiennes* 13 (1978): 105–229.

99. On these sources, see Nicole Bériou, L'avènement des maîtres de la Parole. La prédication à Paris au XIIIe siècle (Paris, 1998), especially chapters 2 and 3.

100. The manuscript contains 243 sermons preached at various chapels in and around Paris during the 1272–1273 liturgical year. On this manuscript, see Bériou, 'La prédication au béguinage de Paris pendant l'année liturgique 1272–1273.' *Recherches augustiennes* 13 (1978): 105–229.This collection and several other sermon manuscripts were housed in the Sorbonne in the thirteenth and fourteenth centuries, making these sermons available to students of the college. See Bériou, *L'avènement des maîtres de la Parole*, 85–8.

101. On these *reportations* and the involvement of clerics affiliated with the Sorbonne, see Bériou *L'avènement des maîtres de la Parole*, chapter 2.
102. 'Vechii les XXXII propriétés de beguinage. Bouche orant, eul plorant, ceur desirant, petit aller, bas regarder, en haut penser, droite entencion, douche pacience, ceur croissant, entendement cherubinal, sentement ceraphinal, aler en seant, parler en taisant, plourer en riant, estre fort en enfleivant, riche en apovriant, sage en taisant, pensees coulees, paroles enmelees, euvres ordenees, foi enluminee, esperance eslevee, amour embrasee, angelique entendement, courtoisie espirituel, devins sentemens, dormir en vellant, vellier en dormant, morir en vivant, vivre en morant, juner en maignant, maignier en junant' BNF lat. 15972, f. 177v.
103. 'Nota quod nullus est bonus predicator vel confessor nisi velit sustine beginagium' BNF 15972, f. 176v.
104. BNF 15972, f. 176r.
105. '[E]t quia vos multa specula in sacra scriptura vidistis et ibi respexistis et affectum vostrum retardastis' BNF 15972, f. 176v.
106. On male intellectuals using female personifications or 'goddesses' in their intellectual endeavors, see Barbara Newman, *God and the Goddesses: Vision, Poetry, and Belief in the Middle Ages* (Philadelphia, 2003).
107. For the relationship between positive and negative representations of beguines in medieval Paris, see Stabler, 'What's in a Name.'

Index